STECK-VAUGHN

Working With Numbers

Consumer Math

AUTHOR

James T. Shea

CONSULTANT

Susan L. Beutel
Consulting/Resource Teacher
Lamoille North Supervisory Union, VT

ACKNOWLEDGMENTS

Executive Editor: Wendy Whitnah
Senior Math Editor: Donna Rodgers
Design Coordinator: John Harrison
Project Design
and Development: The Wheetley Company
Cover Design and
Electronic Production: John Harrison, Adolph Gonzalez and
Chuck Joseph

WORKING WITH NUMBERS SERIES:

Level A	Level D	Consumer Math
Level B	Level E	Refresher
Level C	Level F	Algebra

ISBN: 0-8114-5223-9

1 2 3 4 5 6 7 8 9 0 PO 99 98 97 96 95

TABLE OF CONTENTS

TO THE LEARNER

This book provides you with lessons that review basic skills and other lessons that will help you meet the challenges faced by most consumers. The first two chapters of this book contain lessons on the basic skills of computation needed by the consumer. The remaining chapters provide lessons on how to solve the everyday mathematical problems of the consumer.

This book contains a number of features that will help you master the concepts and skills of consumer mathematics. Among these features are the following.

A Pretest and a Mastery Test

- At the beginning of the book, the Pretest will tell you what material you may know already and what material you have not yet mastered.
- At the end of the book, the Mastery Test will tell you what material you have learned and what you may need to review further.

Unit Reviews

- Each unit concludes with a Unit Review that provides you with an opportunity to demonstrate your understanding of the skills and concepts presented in that particular unit.

Problem-Solving Strategies

- A problem-solving strategy is an effective plan for solving a problem. There are two lessons in each unit on problem-solving strategies. Thus, by the time you finish the book, you will have learned fourteen different strategies for solving problems.

Calculator Applications

- Calculators are an aid to help you with problems that involve lengthy and complicated computations. Each unit contains an application lesson that shows you how to use a calculator as an aid to help you when such problems occur in everyday life.

Answers and Some Solutions

- The answers to all of the problems are provided in the back of the book. This allows you to check your work after it is completed.
- Solutions, or step-by-step explanations, are provided for selected problems. These take you through the steps used to solve the problems.

Name ______________________ Date ______________

PRETEST

Round to the nearest ten.

	a	b	c	d	e
1.	28 ______	92 ______	567 ______	3,249 ______	12,594 ______

Round to the nearest tenth.

	a	b	c	d	e
2.	8.42 ______	25.76 ______	6.184 ______	53.248 ______	24.062 ______

Estimate by rounding both numbers.

	a	b	c	d
3.	726 → × 52 →	59 → × 84 →	4.3 → × 7.2 →	52.1 → × 0.49 →
4.	86) 716 →	43) 358 →	0.26) 2.95 →	5.8) 4.19 →

Find each answer.

	a	b	c	d	e
5.	746 + 8,970	$48.70 + 3.92	9 + 8.36	2,984 3,621 + 975	0.05 5.55 + 55.05
6.	46,795 − 38,952	$91.80 − 8.79	32 − 8.417	40,000 − 6,899	8705 − 827
7.	45 × 36	7.53 × 3	$8.31 × 28	5.7 × 0.4	$57.09 × 46
8.	45) 2,360	5) $35.45	5) 3.01	1.2) 38.52	0.62) 3

Name ____________________ Date ____________

Simplify.

	a	*b*	*c*	*d*
9.	$\frac{12}{21} =$	$\frac{20}{28} =$	$\frac{24}{36} =$	$\frac{30}{40} =$

Find the lowest common denominator (LCD) for each pair of fractions. Then rewrite each pair of fractions as equivalent fractions in higher terms.

	a	*b*	*c*	*d*
10.	$\frac{2}{3} =$ $\frac{5}{6} =$	$\frac{3}{4} =$ $\frac{1}{3} =$	$\frac{5}{8} =$ $\frac{5}{6} =$	$\frac{5}{6} =$ $\frac{7}{9} =$

Change each measurement.

	a	*b*	*c*
11.	23 m = ______ cm	46 kg = ______ g	5.4 L = ______ mL
12.	5 lb = ______ oz	29 in. = ______ ft ______ in.	30 pt = ______ c

Find each answer. Simplify.

	a	*b*	*c*	*d*
13.	$\frac{1}{9} + \frac{5}{6}$	$3\frac{1}{4} + 6\frac{5}{12}$	$7 + 3\frac{5}{8}$	$2\frac{5}{12} + 4\frac{5}{6}$
14.	$\frac{3}{4} - \frac{1}{3}$	$7 - \frac{5}{8}$	$12 - 5\frac{2}{5}$	$8\frac{2}{3} - 4\frac{3}{4}$
15.	$4 \times \frac{5}{8}$	$\frac{6}{7} \times \frac{7}{9}$	$4\frac{1}{2} \times \frac{4}{9}$	$\frac{4}{5} \times 3\frac{3}{4}$
16.	$\frac{2}{3} \div 9$	$\frac{7}{8} \div \frac{1}{4}$	$2\frac{1}{4} \div \frac{9}{10}$	$4\frac{2}{3} \div 2\frac{3}{7}$

Name ______________________ Date ______________

Find each number.

	a	*b*
17.	80% of 250	What percent of 300 is 135?
18.	5% of what number is 4?	65% of 200

Solve.

19. Jamal earns $340.00 for a 40-hour work week. He earns $12.75 for each hour of overtime. How much does he earn for a 45-hour week?

Answer ______________

20. Ghadda is a word processor who earns a weekly salary of $580. What is her yearly salary?

Answer ______________

21. Peter sells medical supplies. He earns a salary of $242 per week plus a commission of 6% on all sales over $3000. One week he had sales of $8900. How much did he earn that week?

Answer ______________

22. In one pay period, Marisa had $621.50 in regular earnings. Her employer withheld $157.54 for taxes and insurance. What was Marisa's take-home pay for that week?

Answer ______________

23. Livia's disability insurance pays 60% of her weekly salary for up to 26 weeks if she is disabled. If her weekly salary is $880, and she is disabled for 15 weeks, how much disability pay will she get?

Answer ______________

24. A store owner pays $30.00 for a jacket. The markup is 70%. For what price should the jacket be sold?

Answer ______________

25. Manuel has a yearly income of $56,000. He budgets 40% for food and clothes. What yearly amount does he budget for food and clothes?

Answer ______________

26. Susan has a balance of $511.26 in her checking account. If she makes a deposit of $152.90 and writes a check for $95.20, what will her new balance be?

Answer ______________

27. Donna has $5600 in a 1-year CD at 2.4% interest compounded annually. How much interest will she receive at the end of 1 year?

Answer ______________

28. A garage mechanic borrowed $5600 for 4 months at 9% interest. How much interest will the mechanic pay on the loan?

Answer ______________

Name ______________________________ Date ______________

29. Athena borrowed $2400 to buy computer equipment. She will repay the loan in 12 installments of $208.80 each. How much interest will she pay on the loan?

Answer ______________

30. If the finance rate on Matt's credit card is 1.5% per month, what is the monthly finance charge on an unpaid balance of $800?

Answer ______________

31. Find the sticker price on a new car that has a base price of $14,822, a destination charge of $378, and these options:

Air Conditioning	$699
CD Player	$349

Answer ______________

32. Bob rented a compact car for $29 per day and drove 1200 miles over a 3-day period. He was allowed 150 free miles per day, and was charged $0.20 for each additional mile. What was the total rental cost?

Answer ______________

33. Ruby has a driver-rating factor of 1.2. The yearly base premium for her automobile insurance is $190.70. How much is her yearly premium?

Answer ______________

34. Suppose gasoline costs $1.18 per gallon and you use 30 gallons per week. How much would you spend for gasoline each week?

Answer ______________

35. If the value of a car depreciates $6000 in one year, what is the average depreciation per month for that year?

Answer ______________

36. Diego rides a train to and from work 22 days each month. A round-trip ticket costs $12.80. A monthly pass costs $240. How much will he save a month by buying a monthly pass?

Answer ______________

37. Leni buys a house for $141,000 and makes a down payment of $60,000. The mortgage payment is $7.34 per thousand dollars of the loan amount. What is her monthly payment?

Answer ______________

38. Lucinda's electric meter reading was 7246 kilowatt hours last month. This month the reading was 7721 kilowatt hours. At 6.5¢ per kilowatt hour, what will her electric bill be for the month?

Answer ______________

Name ____________________ Date ____________

39. Suzanne and George want to insure their condominium for $92,000. The insurance rate is $6.25 per $1000. What will be their annual premium?

Answer ____________

40. Estimate the cost of roofing a building that has 4000 square feet of roof. Shingles cost $35 per square of shingles, and the labor charge is $50 per square. (One square of shingles covers 100 square feet of roof.)

Answer ____________

41. Darren hired a plumber to install new faucets. The basic service charge was $45 for the first hour and $8.50 for each additional 15 minutes. If the job took $1\frac{1}{2}$ hours, what was the cost?

Answer ____________

42. By paying cash instead of charging, Ms. Swann bought a $550 chair for $440. What percent did she save?

Answer ____________

43. Mrs. High's weekly salary is $892. How much is withheld from her pay each week for social security tax? Use a social security tax rate of 7.65%.

Answer ____________

44. Yoshi is filing tax form 1040EZ. Her adjusted gross income was $16,462. She takes a $3700 standard deduction and a $2350 personal exemption. What is her taxable income? If the tax rate is 15%, what is Yoshi's tax liability?

Answer ____________

45. During the past year, $1583 of Danny's pay was withheld for federal income tax. If his tax liability is $1604, will he owe or get a refund? How much?

Answer ____________

46. The market value of a home is $95,000 and the rate of assessment is 40%. The tax rate is 38 mills, or $38 per $1000 of assessed value. What is the yearly real estate tax?

Answer ____________

47. Mr. Homesany is self-employed and made a profit of $26,431 last year. He must pay $3964 in income tax. He also must pay social security tax of 15.3% of his profit. What is Mr. Homesany's total tax liability? Round to the nearest dollar.

Answer ____________

48. Mrs. Bonner bought three computer programs for $39.99, $24.99, and $69.99. The sales-tax rate was 6%. How much sales tax did she pay on the three programs and what was the total cost?

Answer ____________

WHERE TO GO FOR HELP

The table below lists the problems in the Pretest and the pages of the book on which the corresponding skills and concepts are taught and practiced. For each problem that you could not answer or answered incorrectly, you can use the table to find the page number or numbers where that skill or concept is taught.

PROBLEMS	PAGES	PROBLEMS	PAGES	PROBLEMS	PAGES	PROBLEMS	PAGES
1a	14	6d	15	13a	35	25	85
1b	14	6e	15	13b	36	26	87
1c	14	7a	18	13c	36	27	90
1d	14	7b	19	13d	36	28	97
1e	14	7c	19	14a	37	29	101
2a	14	7d	19	14b	38	30	102
2b	14	7e	19	14c	38	31	104
2c	14	8a	22	14d	38	32	107
2d	14	8b	23	15a	42	33	111
2e	14	8c	23	15b	42	34	112
3a	24	8d	23	15c	42	35	117
3b	24	8e	23	15d	42	36	120
3c	24	9a	32	16a	43	37	128
3d	24	9b	32	16b	43	38	131
4a	24	9c	32	16c	43	39	132
4b	24	9d	32	16d	43	40	140
4c	24	10a	32	17a	48–49	41	141
4d	24	10b	32	17b	51	42	146
5a	15	10c	32	18a	50	43	150
5b	16	10d	32	18b	48–49	44	152–154
5c	16	11a	27	19	59	45	152–156
5d	15	11b	27	20	60	46	165
5e	16	11c	27	21	63	47	180
6a	15	12a	34	22	66	48	170
6b	16	12b	34	23	76		
6c	16	12c	34	24	71		

Reading and Writing Numbers

A place-value chart can help you read and write whole numbers and decimals. Each digit in a number has a value based on its place in the number.

We read and write the number in the place-value chart as: four hundred fifty-two thousand, five hundred twenty-one and six hundred thirteen thousandths.

Notice that commas are used to separate the digits into groups of three. This helps make larger numbers easier to read.

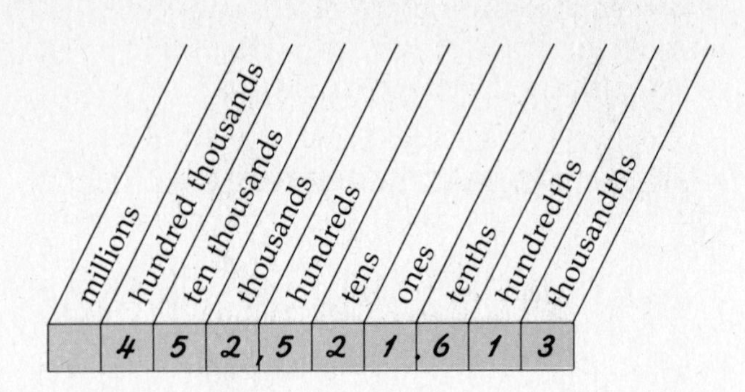

Write each number in the place-value chart.

1. 7,646,918.764
2. 189,498.004
3. 96,950.53
4. $2000.99

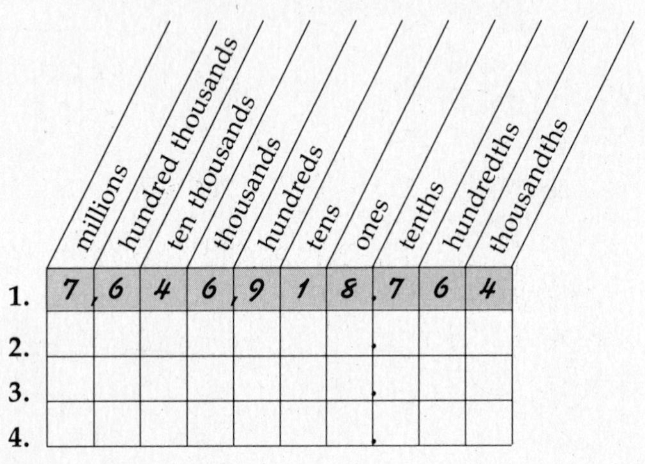

Write each number using digits. Insert commas where needed.

5. ninety-nine and one hundredth ______ 99.01
6. fifty-six thousand, forty-seven and thirty-one hundredths ______
7. forty-two dollars and fifty-three cents ______
8. eight and six hundred twelve thousandths ______

Write each number in words. Insert commas where needed.

9. 50.8 ______ fifty and eight tenths
10. 816.07 ______
11. 385.41 ______

______

12. 1005.075 ______

______

13. $13.99 ______
14. $210.35 ______

REVIEWING BASIC SKILLS

Comparing and Ordering

To compare two numbers, begin at the left.
Compare the digits in each place.

Remember,

- zeros at the end of a decimal do not change the value of the decimal. 2 = 2.0 = 2.00; 2.2 = 2.20
- you can write zeros so that the numbers have the same number of places.

Compare. Write <, >, or =.

	a	*b*	*c*
1.	58 __>__ 52	48 ______ 48.0	88 ______ 62
	5 8 5 2	4 8 4 8 . 0	8 8 6 2
2.	183 ______ 95	21 ______ 215	37 ______ 37.0
	1 8 3 9 5	2 1 2 1 5	3 7 3 7 . 0
3.	0.789 ______ 0.7	2399.9 ______ 2400.0	9.8 ______ 9.20
4.	23.56 ______ 28	689 ______ 689.1	345.987 ______ 895.67

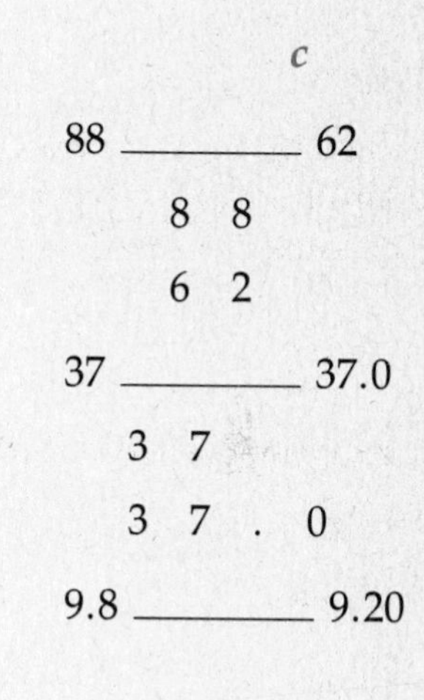

Write in order from least to greatest.

5. 76 45 98 ____45 76 98____

6. 789 287 891 ______________________

7. 190 754 561 ______________________

8. 3963 9867 1256 ______________________

9. 8432 2876 8765 ______________________

10. 78.67 84.13 32.85 ____32.85 78.67 84.13____

11. 375.2 973.9 24.678 ______________________

12. 0.674 0.340 0.970 ______________________

13. 12.976 12.975 12.97 ______________________

14. 7.6 4.5 9.8 ______________________

15. 190 1.90 0.190 ______________________

REVIEWING BASIC SKILLS

Rounding

Often you will need to round numbers to estimate or to make a problem easier to understand.

Remember, 5 or greater rounds up.

Round to the nearest ten.

	a	b	c	d
1.	45 50	87 ____	93 ____	21 ____
2.	478 ____	311 ____	3255 ____	4189 ____

Round to the nearest hundred.

	a	b	c	d
3.	521 500	670 ____	427 ____	752 ____
4.	4350 ____	5375 ____	11,755 ____	68,906 ____

Round to the nearest tenth.

	a	b	c	d
5.	9.34 9.3	7.27 ____	3.55 ____	45.67 ____
6.	1.111 ____	4.932 ____	22.760 ____	39.129 ____

Round to the nearest hundredth.

	a	b	c	d
7.	2.831 2.83	15.947 ____	6.295 ____	81.608 ____
8.	4.9023 ____	0.6105 ____	1.8264 ____	0.0995 ____

Round to the nearest dollar.

	a	b	c	d
9.	$24.25 $24	$24.75 ____	$39.99 ____	$95.95 ____
10.	$329.87 ____	$836.49 ____	$1120.95 ____	$1999.95 ____

REVIEWING BASIC SKILLS

Addition and Subtraction of Whole Numbers

Add. Regroup when necessary.

	a	b	c	d	e
1.	474 +232 706	305 + 76	148 +516	26 +781	950 +394
2.	892 +8902	65,987 +39,760	12,896 +90,642	6540 + 361	43,179 + 5,087
3.	7893 8093 +1563	98,453 4,721 +75,230	7,543 59,932 + 6,081	839 12,476 + 5,390	72,166 205 +31,654

Subtract. Regroup when necessary.

	a	b	c	d	e
4.	764 −156 608	562 − 48	8291 − 783	1936 − 479	4301 −2999
5.	45,000 −34,029	54,764 − 6,987	90,760 −39,263	119,076 −118,999	13,821 − 4,296
6.	72,541 − 5,316	84,205 −37,009	92,640 − 8,721	50,013 − 6,835	10,000 − 4,999

Line up the digits. Then add or subtract. Regroup when necessary.

	a	b	c
7.	6573 + 34 = ______ 6573 + 34	746 − 347 = ______	46,527 + 547,907 = ______

REVIEWING BASIC SKILLS

Addition and Subtraction of Decimals

Add. Write zeros as needed.

	a	*b*	*c*	*d*	*e*
1.	7.20 +5.47 12.67	60.9 +37.6	15.4 +93.2	8 +1.7	6.7 +1.9
2.	\$3.48 + 0.49	\$892.10 + 1.73	\$12.92 + 67.02	\$10.66 + 19.29	\$1.99 + 0.57
3.	75.93 401.3 +1561	7845.3 4462.1 +8752.27	397.543 269.31 +3135.8	17.431 90.459 + 8.02	63.95 0.08 +9

Subtract. Write zeros as needed.

	a	*b*	*c*	*d*	*e*
4.	7 10 8.0 −6.6 1.4	7.24 −3.47	8.9 −4	6.15 −0.09	12 − 5.1
5.	\$892.10 − 0.78	\$2600.75 − 340.29	\$557.63 − 6.98	\$190.76 − 14.99	\$300 − 37.58
6.	9056.3 − 32.76	3.729 −0.086	8.4 −2.173	62.143 − 9.5	4.01 −2.107

Line up the digits. Write zeros as needed. Then add or subtract.

	a	*b*	*c*
7.	65.73 − 0.34 = ______ 65.73 − 0.34	734 − 44.7 = ______	709.8 + 807.87 = ______

REVIEWING BASIC SKILLS

Estimation of Sums and Differences

Estimate the sum or difference to the nearest ten.

	a	b	c	d
1.	139 → 140 + 53 → + 50 190	821 → +9853 →	6789 → −2345 →	129,201 → − 45,679 →

Estimate the sum or difference to the nearest hundred.

	a	b	c	d
2.	7878 → 7900 + 621 → + 600 8500	13,542 → +67,890 →	8943 → −6785 →	12,641 → − 9,870 →

Estimate the sum or difference to the nearest one.

	a	b	c	d
3.	$ 1.99 → $ 2 12.19 → 12 + 110.79 → +111 $125	$ 31.47 → 568.76 → + 23.98 →	$57.48 → − 32.40 →	$82.72 → − 64.02 →

Estimate the sum or difference to the nearest tenth.

	a	b	c	d
4.	20.01 → 20.0 +69.29 → +69.3 89.3	50.75 → +73.94 →	8.56 → −6.25 →	3.44 → −0.89 →

Estimate the sum or difference to the nearest hundredth.

	a	b	c	d
5.	2.379 → 2.38 +0.026 → +0.03 2.41	5.342 → +3.876 →	478.287 → −235.61 →	50.621 → − 7.878 →

Multiplication of Whole Numbers

Multiply.

	a	*b*	*c*	*d*	*e*
1.	36 × 2 = 72	21 × 4	401 × 8	6812 × 7	89,920 × 9
2.	52 × 10	42 × 58	59 × 36	547 × 84	4000 × 70
3.	186 × 78	200 × 46	539 × 68	417 × 30	263 × 99
4.	985 × 340	544 × 807	196 × 532	702 × 100	357 × 498

Line up the digits. Then multiply.

	a	*b*	*c*
5.	6573 × 9 = ______	108 × 247 = ______	56 × 92 = ______
6.	500 × 41 = ______	75 × 38 = ______	20 × 96 = ______

6573
× 9

Multiplication of Decimals

Multiply.

	a	*b*	*c*	*d*	*e*
1.	34.1 × 2.5	6.7 × 0.8	19.3 × 4.6	892.5 × 73.1	75.8 × 30.9
2.	1.875 × 2	$3.49 × 16	$68.02 × 53	54,000 × 0.18	259 × 0.07
3.	0.435 × 0.276	0.126 × 0.007	0.009 × 0.362	3.078 × 0.901	0.301 × 1.564

Example (1a): carry 2; 1705; 6820; 85.25

Line up the digits. Then multiply.

	a	*b*	*c*
4.	12.73 × 0.54 = ________	4.45 × 3.9 = ________	64,527 × 0.09 = ________
5.	5.98 × 0.87 = ________	81.7 × 6.52 = ________	29,173 × 0.4 = ________

Example (4a): 12.73 × 0.54

PROBLEM-SOLVING STRATEGY

Choose an Operation

Sometimes a problem does not tell you whether to add, subtract, multiply, or divide. To solve such a problem, you must read the problem carefully. Then, decide what the problem is asking you to do. Next, choose an operation to solve the problem. Finally, solve the problem.

EXAMPLE 1

Read the problem.

Robin bought 2.35 pounds of grapes and 1.59 pounds of plums. How many pounds of fruit did she buy?

Decide what the problem is asking.

In this problem, the question "How many pounds of fruit..." is asking you to find a total or sum.

Choose the operation.

To solve, you must add.

Solve the problem.

$2.35 + 1.59 = 3.94$

Robin bought a total of 3.94 pounds of fruit.

EXAMPLE 2

Read the problem.

A plumber charged $798 for 19 hours of work. The charge was the same for each hour. What was the plumber's hourly rate?

Decide what the problem is asking.

In this problem, the question "What was the ...rate?" is asking you to find equal parts.

Choose the operation.

To solve, you must divide.

Solve the problem.

$\$798 \div 19 = \$42$. The plumber's rate was $42 per hour.

Choose the operation needed to solve each problem. Then solve.

1. In April, the Randolf family spent $483 for food. This was $37 more than the amount spent in March. How much did the Randolf family spend for food in March?

   Operation ______________________

   Answer ______________________

2. In one day, 934 concert tickets were sold for $23 each. How much money was taken in from the sale of the tickets?

   Operation ______________________

   Answer ______________________

Choose the operation needed to solve each problem. Then solve.

3. How far can a bus travel in 4.5 hours if its average speed is 52 miles per hour?

Operation ______________________

Answer ______________________

4. A shoe store sold 1620 pairs of shoes in a 30-day period. What was the average number of pairs sold per day?

Operation ______________________

Answer ______________________

5. To buy a computer, Yvonne is making 36 equal payments of $54 each. What is the total of the payments?

Operation ______________________

Answer ______________________

6. Brenda drove 1329 miles in January and 1893 miles in February. How many more miles did she drive in February?

Operation ______________________

Answer ______________________

7. The payroll of one department in a certain company is $2933. Each of the 7 employees receives the same pay. What is the pay for each employee?

Operation ______________________

Answer ______________________

8. On Monday, Hal's club collected 92.3 pounds of newspapers for recycling. On Tuesday, the amount collected was 24.2 pounds more than on Monday. How much was collected on Tuesday?

Operation ______________________

Answer ______________________

9. Vernon's phone bill was $54.77 in March. In April, the bill was $6.89 less than in March. How much was the phone bill in April?

Operation ______________________

Answer ______________________

10. A card table costs $24.96 and the chairs are $11.84 each. Find the cost of the table and two chairs.

Operation ______________________

Answer ______________________

11. Ernie earned $60.32 for working 8 hours. What was the hourly rate?

Operation ______________________

Answer ______________________

12. Jessica weighs 45.7 kilograms. She weighs 4.2 kilograms more than Jenny. How much does Jenny weigh?

Operation ______________________

Answer ______________________

Division of Whole Numbers

Divide and check.

	a	b	c	d
1.	45)567	23)543	39)303	12)6489
2.	56)785	27)678	76)564	34)8867
3.	64)7865	55)3030	40)8765	60)4320
4.	21)8442	43)4343	62)3723	39)7862

Example (1a):

```
    12 R27
45)567
   45
   117
    90
    27
```

Set up the problem. Then divide and check.

	a	b	c
5.	897 ÷ 2 = ________	389 ÷ 14 = ________	2375 ÷ 32 = ________
6.	5678 ÷ 54 = ________	6352 ÷ 76 = ________	1698 ÷ 12 = ________

Example (5a): 2)897

Division of Decimals

Divide. Round to the nearest tenth if needed. Write zeros as needed.

	a	b	c	d
1.	$6.8\overline{)5.44}$ = 0.8 (5 44, 0)	$3\overline{)21.3}$	$8.7\overline{)8.874}$	$12\overline{)64.44}$
2.	$4.1\overline{)23.616}$	$12.9\overline{)387}$	$4.6\overline{)12.926}$	$9.1\overline{)88.27}$
3.	$0.94\overline{)78.02}$	$0.28\overline{)0.364}$	$0.003\overline{)17.736}$	$0.16\overline{)34.88}$
4.	$78\overline{)7.8}$	$2.7\overline{)0.216}$	$0.65\overline{)12.61}$	$2.6\overline{)0.3354}$

Set up the problem. Divide. Round to the nearest tenth if needed. Write zeros as needed.

	a	b	c
5.	$0.754 \div 5.8 =$ ________	$248.2 \div 34 =$ ________	$84.15 \div 0.45 =$ ________
	$5.8\overline{)0.754}$		
6.	$80.46 \div 2.7 =$ ________	$0.99 \div 9.9 =$ ________	$306.6 \div 1.4 =$ ________

REVIEWING BASIC SKILLS

Estimation of Products and Quotients

Estimate by rounding both numbers.

	a	*b*	*c*	*d*
1.	854×47 → $850 \times 50 = 42{,}500$	361×55 →	12×43 →	921×89 →
2.	0.85×5.90 →	2.1×64 →	0.36×19.8 →	93.6×0.83 →

Estimate using basic facts.

	a	*b*	*c*	*d*
3.	$4\overline{)475}$ → $4\overline{)480}$ = 120	$3\overline{)359}$ →	$7\overline{)703}$ →	$2\overline{)455}$ →
4.	$8\overline{)4.82}$ →	$4\overline{)390}$ →	$6\overline{)23.8}$ →	$5\overline{)1.96}$ →

Estimate by rounding both numbers.

	a	*b*	*c*	*d*
5.	$54\overline{)746}$ → $50\overline{)750}$ = 15	$78\overline{)315}$ →	$15\overline{)602}$ →	$83\overline{)721}$ →
6.	$8.5\overline{)355}$ →	$0.53\overline{)2.97}$ →	$89\overline{)362}$ →	$5.8\overline{)3.64}$ →

REVIEWING BASIC SKILLS

Calculator Applications

A calculator can help you understand place value. Any standard number can be written as hundreds, tens, tenths, hundredths, and so on.

EXAMPLE 1 230 = ______ hundreds

Divide 230 by 100 mentally. Think of moving the decimal point.

230 = <u>2.3</u> hundreds

Check the answer with a calculator.

230 [÷] 100 [=] [2.3]

EXAMPLE 2 5.281 = ______ tenths

Dividing by 0.1 is the same as multiplying by 10. Multiply 5.281 by 10 mentally.

5.281 = <u>52.81</u> tenths

Check by dividing with a calculator.

5.281 [÷] 0.1 [=] [52.81]

EXAMPLE 3 A gram is one-thousandth of a kilogram. How many grams are in 32 kilograms?

32 = ______ thousandths

Dividing by 0.001 is the same as multiplying by 1000. Multiply 32 by 1000 mentally.

32 = <u>32,000</u> thousandths

Check the answer with a calculator. Divide 32 by a thousandth or 0.001.

32 [÷] 0.001 [=] [32000.]

So, 32 kg = 32,000 g.

Compute the answer mentally. Then check by using a calculator.

1. 40.7 = <u>*4.07*</u> tens
2. 40.7 = ______ tenths
3. 40.7 = ______ hundreds
4. 2.84 = ______ hundredths
5. 2.84 = ______ tenths
6. 2.84 = ______ tens
7. 391.2 = ______ hundreds
8. 391.2 = ______ tens
9. 0.84 = ______ hundredths
10. 0.84 = ______ tenths
11. 3401 = ______ tenths
12. 450 = ______ thousands

Solve. Use the method shown above.

13. A centimeter is one-hundredth of a meter. How many centimeters are in 20 meters?

20 = <u>*2000*</u> *hundredths*

Answer ______________

14. A millimeter is one-thousandth of a meter. How many millimeters are in 43.2 meters?

Answer ______________

15. A milligram is one-thousandth of a gram. How many milligrams are in 3.7 grams?

Answer ______________

16. A kilogram is one thousand grams. How many kilograms are in 734 grams?

Answer ______________

PROBLEM SOLVING

Applications

Solve.

1. A passenger train has 14 coaches. The length of each coach is 59 ft. What is the total length of the train without the engine?

Answer ____________________

2. A bushel of potatoes weighs about 26 kilograms. What is the weight of 5225 bushels loaded into a boxcar for shipment?

Answer ____________________

3. The Garcias own a citrus grove. They pay $6.45 per acre each time they have it irrigated. If the grove contains 48 acres, what does it cost each time for irrigating?

Answer ____________________

4. John could jump about 6 times as high on the moon as he could on Earth. If he could jump 30.6 feet on the moon how high could he jump on Earth?

Answer ____________________

5. An airliner carried 73 passengers from San Francisco to Chicago. The fare was $239.00. How much money was received by the airline for that flight?

Answer ____________________

6. An agriculture class raised 2535 bushels of potatoes on 39 acres. What was the average yield per acre?

Answer ____________________

7. Mrs. George made one payment of $37.50 and one payment of $73.94. How much did she pay in all?

Answer ____________________

8. A jet plane made a flight of 6510 miles in 15.5 hours. What average speed did it make?

Answer ____________________

Metric Measurement

The basic metric units are meter (m), liter (L), and gram (g). These basic unit names are used with prefixes to form the names of larger and smaller units.

To change from one unit to another, find the unit in the chart below. Move to the new unit. When you move right to a smaller unit, multiply by 10 for each step. When you move left to a larger unit, divide by 10 for each step.

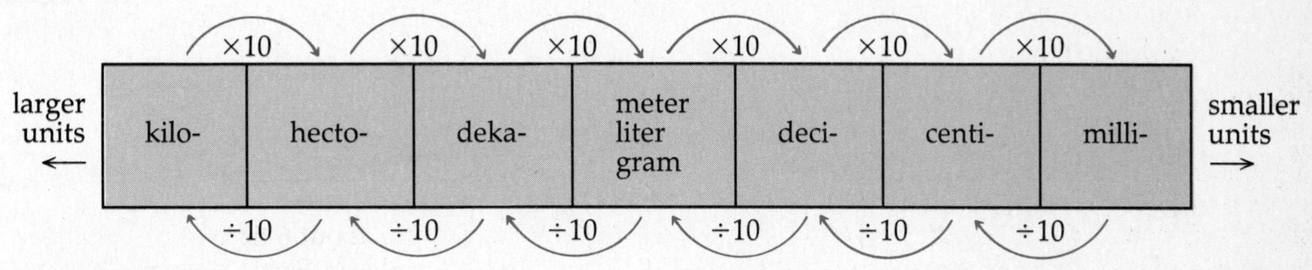

Find: 15 hm = ________ m

To change hm to m, find hecto in the table. Find meter in the table. To change hm to a smaller unit, multiply by 10 for each step.

15 hm = (15 × 10 × 10) m
= (15 × 100) m
15 hm = 1500 m

You will have more smaller units.

Find: 9000 mL = ________ L

To change mL to L, find milli in the table. Find liter in the table. To change mL to a larger unit, divide by 10 for each step.

9000 mL = (9000 ÷ 10 ÷ 10 ÷ 10) L
= (9000 ÷ 1000) L
9000 mL = 9 L

You will have fewer larger units.

Change each measurement to a smaller unit.

	a	*b*	*c*
1.	5 g = ________ mg	127 L = ________ dL	97 kg = ________ g
2.	46 cm = ________ mm	12 km = ________ m	8 L = ________ mL

Change each measurement to a larger unit.

	a	*b*	*c*
3.	48,000 L = ________ kL	980 mm = ________ cm	2100 mg = ________ g
4.	500 hg = ________ kg	17,000 m = ________ km	3000 g = ________ kg

PROBLEM-SOLVING STRATEGY

Identify Extra Information

Some problems contain information that is not needed to solve the problem. When you read a problem, it may be helpful to cross out the extra facts.

Read the problem.

Everett typed a 500-word essay in 6 minutes 30 seconds. How many seconds did it take him to type the essay?

Decide which facts are needed.

The problem asks for the number of seconds. You need to change 6 minutes to seconds, then add 30 seconds. Recall that there are 60 seconds in a minute.

Decide which facts are extra.

You do not need to know that the essay has 500 words.

Solve the problem.

Multiply the number of minutes by 60. Then add 30 seconds.

Seconds in 6 minutes: $6 \times 60 = 360$ seconds

Total seconds: $360 + 30 = 390$ seconds

Everett typed the essay in 390 seconds.

In each problem, cross out the extra facts. Then solve the problem.

1. David worked 8.5 hours on Monday, 6 hours on Tuesday, ~~and 7.5 hours on Wednesday.~~ How much longer did he work on Monday than Tuesday?

Answer ____________________

2. Winona bought a car that cost $9880, including a $199 service warranty. She paid $1000 as a down payment and took out a loan for the rest. What was the amount of her loan?

Answer ____________________

3. Raul works part-time and earns $75 per week. If he saves about $25 per week, how much money will he save in 6 weeks?

Answer ____________________

4. Kris spent $4.32 for a calculator, $12.89 for a backpack, and $8.99 for a lantern. How much more did the lantern cost than the calculator?

Answer ____________________

In each problem, cross out the extra facts. Then solve the problem.

5. A ream of paper contains 500 sheets. If a person buys 12 reams for $5.25 each, what is the total cost?

Answer ______________________

6. A 16-inch pizza costs $13.95. If five friends share the cost equally, how much does each friend pay?

Answer ______________________

7. The original price of a jogging suit was $38.50. Tommy bought the suit on sale for $28.95. He also spent $6.29 for gloves. How much did Tommy spend for the jogging suit and gloves?

Answer ______________________

8. Martha is 63 inches tall and weighs 115 pounds. Her sister is 67 inches tall and weighs 122 pounds. How much shorter is Martha than her sister?

Answer ______________________

9. Jaleesa paid $2.69 for an 8-foot strip of molding. If she cut off 2 feet 4 inches of molding, find the length of the remaining piece.

Answer ______________________

10. The distance from Greg's house to school is 4.2 kilometers. The distance to the library is 2.7 kilometers. If Greg rides his bike from home to school and back, how far does he ride?

Answer ______________________

11. Evelyn worked 4 hours 15 minutes before lunch, then took a break for 45 minutes. She worked 3 hours 30 minutes after lunch. How much time was she working that day?

Answer ______________________

12. Ms. Benitez works part-time and earns $44 per week. She earns $5.50 per hour. How much does she earn in 12 weeks?

Answer ______________________

13. Brad drove 134 miles in 2 hours 20 minutes. How many minutes did he drive?

Answer ______________________

14. A carton of milk contains 250 milliliters and costs 39¢. How many cartons could be filled from 4 liters of milk?

Answer ______________________

PROBLEM SOLVING

Applications

Solve.

1. Joe went to the grocery story and purchased meat for $4.48, butter for $1.88, tomatoes for $0.87, and canned goods for $3.89. How much did all of these cost?

Answer ______________________

2. Mr. Burt bought gas for the car at $9.75, oil for $0.99, a new tire for $68.75, and a filter for $5.15. What was the total bill?

Answer ______________________

3. Elida gave the storekeeper a twenty-dollar bill for a purchase of $14.51. How much change should she have received?

Answer ______________________

4. Maria bought 4 yards of material at $3.52 a yard. How much did she pay for the 4 yards?

Answer ______________________

5. Jorge bought 3 shirts at $18.49 each. How much did he pay for the 3 shirts?

Answer ______________________

6. Mrs. Turner paid $7.47 for a three-pound roast. What was the price per pound?

Answer ______________________

7. Mr. Parker drove 720 kilometers in 9 hours. What was his average rate of speed?

Answer ______________________

8. Josie bought five oranges for $1.15. At this price, how much would one orange cost?

Answer ______________________

Unit 1 Review

Add or subtract.

	a	*b*	*c*	*d*
1.	343 + 23	452.9 − 342.8	55,234 − 2,566	\$209.89 + 178.93
2.	78.3 + 0.23	3 − 2.6	6271 − 4286	276.0 + 15.9
3.	45,493 − 7,723	6264 + 8740	6.234 − 5.667	9.89 + 203.9 + 648.9

Multiply or divide.

	a	*b*	*c*	*d*
4.	73 × 25	2.93 × 0.81	32,234 × 6	439.89 × 0.9
5.	$8\overline{)8976}$	$76\overline{)8.74}$	$0.97\overline{)67{,}221}$	$5.2\overline{)5428.8}$

Estimate.

	a	*b*	*c*
6.	65.90 → + 553.6 →	38.5 − 23	145 + 765
7.	387 × 6	$97\overline{)436}$	258 × 3.5

Change each measurement.

	a	*b*	*c*
8.	37 m = ______ cm	397 kg = ______ g	1.9 L = ______ mL

UNIT 2

REVIEWING BASIC SKILLS

Fractions

Write the fraction for each word name.

	a	*b*	*c*
1.	two fifths $\frac{2}{5}$	one ninth ________	five sixtieths ________

Write the word name for each fraction.

	a	*b*	*c*
2.	$\frac{2}{3}$ *two thirds*	$\frac{7}{10}$ ________	$\frac{1}{88}$ ________

There are 12 eggs in a dozen. Each egg is $\frac{1}{12}$ dozen. Write the following as a fraction of a dozen. Write the word name for the fraction.

	a	*b*	*c*
3.	7 eggs = $\frac{7}{12}$ dozen	5 eggs = ________ dozen	11 eggs = ________ dozen
	seven twelfths	________	________

Simplify.

	a	*b*	*c*	*d*
4.	$\frac{12}{18} = \frac{2}{3}$ $\frac{12 \div 6}{18 \div 6} = \frac{2}{3}$	$\frac{3}{6} =$ ________	$\frac{10}{15} =$ ________	$\frac{8}{10} =$ ________
5.	$\frac{6}{6} =$ ________	$\frac{8}{16} =$ ________	$\frac{6}{9} =$ ________	$\frac{12}{14} =$ ________

Find the lowest common denominator (LCD) for each pair of fractions. Then rewrite each pair of fractions as equivalent fractions in higher terms.

	a	*b*	*c*	*d*
6.	$\frac{3}{4} = \frac{3 \times 5}{4 \times 5} = \frac{15}{20}$	$\frac{1}{2} =$	$\frac{7}{10} =$	$\frac{11}{12} =$
	$\frac{1}{5} = \frac{1 \times 4}{5 \times 4} = \frac{4}{20}$	$\frac{2}{3} =$	$\frac{2}{15} =$	$\frac{5}{8} =$

REVIEWING BASIC SKILLS

Mixed Numbers

Write as a mixed number or whole number. Simplify.

	a	*b*	*c*
1.	$\frac{15}{15}$ = 1 $\quad 15\overline{)15}$ (quotient 1)	$\frac{6}{6}$ = ____	$\frac{9}{9}$ = ____
2.	$\frac{27}{9}$ = ____	$\frac{32}{8}$ = ____	$\frac{36}{6}$ = ____
3.	$\frac{37}{5}$ = $7\frac{2}{5}$ $\quad 5\overline{)37}$ (quotient $7\frac{2}{5}$), 35, 2	$\frac{24}{10}$ = ____	$\frac{54}{7}$ = ____
4.	$\frac{59}{8}$ = ____	$\frac{31}{8}$ = ____	$\frac{27}{4}$ = ____
5.	$\frac{73}{9}$ = ____	$\frac{145}{16}$ = ____	$\frac{107}{9}$ = ____

Write as an improper fraction.

	a	*b*	*c*
6.	3 = $\frac{3}{1}$	7 = ____	2 = ____
7.	$5\frac{3}{5}$ = $\frac{28}{5}$ $5 \times 5 + 3 = 28$	$4\frac{2}{7}$ = ____	$14\frac{6}{11}$ = ____
8.	$8\frac{2}{11}$ = ____	$24\frac{6}{7}$ = ____	$4\frac{3}{10}$ = ____
9.	$1\frac{7}{9}$ = ____	$3\frac{1}{7}$ = ____	$15\frac{1}{5}$ = ____
10.	$6\frac{2}{9}$ = ____	$5\frac{3}{7}$ = ____	$1\frac{5}{9}$ = ____

Customary Measurement

Some customary units of length are the inch, foot, yard, and mile. The chart shows the relationship of one unit to another.

1 foot (ft)	= 12 inches (in.)
1 yard (yd)	= 3 ft
	= 36 in.
1 mile (mi)	= 1760 yd
	= 5280 ft

Some customary units of weight are the ounce, pound, and ton. The chart shows the relationship of one unit to another.

1 pound (lb)	= 16 ounces (oz)
1 ton (T)	= 2000 pounds

Some customary units of capacity are the cup, pint, quart, and gallon. The chart shows the relationship of one unit to another.

1 pint (pt)	= 2 cups (c)
1 quart (qt)	= 2 pt
	= 4 c
1 gallon (gal)	= 4 qt
	= 8 pt
	= 16 c

Change each measurement to a larger unit.

	a	*b*
1.	98 in. = 8 ft 2 in.	37 in. = ____ ft ____ in.
2.	28 oz = ____ lb ____ oz	64 oz = ____ lb ____ oz
3.	9 pt = ____ qt ____ pt	31 qt = ____ gal ____ qt

$$12\overline{)98} = 8 \text{ R } 2 \quad (98 - 96 = 2)$$

Change each measurement to a smaller unit.

	a	*b*	*c*
4.	2 yd = 72 in.	15 ft = ________ in.	2 mi = ________ ft
5.	3 lb = ________ oz	20 lb = ________ oz	14 T = ________ lb
6.	8 qt 1 c = ________ c	10 pt = ________ c	4 gal = ________ pt

$2 \times 36 = 72$

Addition of Fractions

Remember,

- to simplify a proper fraction, write it in simplest terms.
- to simplify an improper fraction, write it as a whole or mixed number.

Add. Simplify.

	a	*b*	*c*	*d*
1.	$\frac{4}{7} + \frac{3}{7}$ $\frac{7}{7} = 1$	$\frac{10}{13} + \frac{4}{13}$	$\frac{2}{3} + \frac{1}{3}$	$\frac{12}{17} + \frac{1}{17}$
2.	$\frac{7}{11} = \frac{14}{22}$ $+\frac{4}{22} = \frac{4}{22}$ $\frac{18}{22} = \frac{9}{11}$	$\frac{2}{5} + \frac{1}{3}$	$\frac{4}{8} + \frac{1}{4}$	$\frac{8}{11} + \frac{2}{9}$
3.	$\frac{6}{7} = \frac{12}{14}$ $+\frac{1}{2} = \frac{7}{14}$ $\frac{19}{14} = 1\frac{5}{14}$	$\frac{3}{5} + \frac{2}{3}$	$\frac{9}{12} + \frac{3}{4}$	$\frac{1}{2} + \frac{2}{3}$
4.	$\frac{5}{8} = \frac{15}{24}$ $+\frac{1}{3} = \frac{8}{24}$ $\frac{23}{24}$	$\frac{1}{4} + \frac{1}{3}$	$\frac{5}{6} + \frac{3}{4}$	$\frac{3}{5} + \frac{2}{3}$

Line up the digits. Add and simplify.

	a	*b*	*c*
5.	$\frac{2}{7} + \frac{2}{6} =$ ________ $\frac{2}{7} + \frac{2}{6}$	$\frac{2}{3} + \frac{4}{5} =$ ________	$\frac{7}{9} + \frac{6}{8} =$ ________
6.	$\frac{4}{5} + \frac{2}{4} =$ ________	$\frac{3}{9} + \frac{9}{12} =$ ________	$\frac{7}{9} + \frac{2}{10} =$ ________

REVIEWING BASIC SKILLS

Addition of Mixed Numbers

Add. Simplify.

	a	*b*	*c*	*d*
1.	$4\frac{4}{12} + 3\frac{7}{12}$ $7\frac{11}{12}$	$5\frac{3}{4} + 1\frac{2}{4}$	$3\frac{5}{6} + 7\frac{3}{6}$	$3\frac{9}{10} + 1\frac{7}{10}$
2.	$8\frac{2}{6} = 8\frac{4}{12}$ $+3\frac{4}{12} = 3\frac{4}{12}$ $11\frac{8}{12} = 11\frac{2}{3}$	$7\frac{3}{5} + 2\frac{5}{10}$	$23\frac{2}{7} + 16\frac{1}{14}$	$13\frac{3}{8} + 21\frac{3}{4}$
3.	$6\frac{8}{9} = 6\frac{8}{9}$ $+7\frac{2}{3} = 7\frac{6}{9}$ $13\frac{14}{9} = 14\frac{5}{9}$	$32\frac{4}{5} + 14\frac{6}{15}$	$4\frac{6}{8} + 25\frac{1}{2}$	$16\frac{4}{32} + 34\frac{4}{8}$
4.	$3\frac{4}{5} = 3\frac{16}{20}$ $+2\frac{3}{4} = 2\frac{15}{20}$ $5\frac{31}{20} = 6\frac{11}{20}$	$14\frac{5}{6} + 34\frac{8}{9}$	$4\frac{2}{7} + 2\frac{1}{3}$	$14\frac{1}{3} + 15\frac{6}{8}$
5.	$7\frac{2}{5} = 7\frac{6}{15}$ $+5\frac{1}{3} = 5\frac{5}{15}$ $12\frac{11}{15}$	$7\frac{2}{9} + 3\frac{3}{5}$	$5\frac{5}{6} + 6\frac{2}{8}$	$25\frac{5}{8} + 67\frac{3}{5}$

Line up the digits. Add. Simplify.

	a	*b*	*c*
6.	$2\frac{3}{8} + 7\frac{3}{5} =$ ________ $2\frac{3}{8}$ $+7\frac{3}{5}$	$9\frac{8}{11} + 3\frac{4}{9} =$ ________	$14\frac{2}{5} + 16\frac{2}{3} =$ ________

REVIEWING BASIC SKILLS

Subtraction of Fractions

Subtract. Simplify.

	a	*b*	*c*	*d*
1.	$\frac{11}{12} - \frac{7}{12}$ $\frac{4}{12} = \frac{1}{3}$	$\frac{7}{9} - \frac{5}{9}$	$\frac{11}{15} - \frac{8}{15}$	$\frac{5}{8} - \frac{3}{8}$
2.	$\frac{5}{6} = \frac{10}{12}$ $-\frac{7}{12} = \frac{7}{12}$ $\frac{3}{12} = \frac{1}{4}$	$\frac{7}{8} - \frac{3}{16}$	$\frac{6}{7} - \frac{3}{14}$	$\frac{3}{8} - \frac{1}{4}$
3.	$\frac{7}{8} = \frac{7}{8}$ $-\frac{1}{2} = \frac{4}{8}$ $\frac{3}{8}$	$\frac{4}{5} - \frac{4}{15}$	$\frac{8}{9} - \frac{1}{3}$	$\frac{5}{6} - \frac{1}{18}$
4.	$\frac{5}{7} = \frac{15}{21}$ $-\frac{1}{3} = \frac{7}{21}$ $\frac{8}{21}$	$\frac{7}{9} - \frac{1}{5}$	$\frac{5}{6} - \frac{1}{4}$	$\frac{5}{8} - \frac{1}{6}$
5.	$\frac{4}{9} = \frac{20}{45}$ $-\frac{2}{5} = \frac{18}{45}$ $\frac{2}{45}$	$\frac{1}{2} - \frac{2}{9}$	$\frac{5}{8} - \frac{4}{7}$	$\frac{1}{2} - \frac{2}{7}$

Line up the digits. Subtract. Simplify.

	a	*b*	*c*
6.	$\frac{9}{11} - \frac{2}{3} =$ ________ $\frac{9}{11}$ $-\frac{2}{3}$	$\frac{3}{5} - \frac{1}{2} =$ ________	$\frac{8}{9} - \frac{3}{4} =$ ________

REVIEWING BASIC SKILLS

Subtraction of Mixed Numbers

Subtract. Simplify.

	a	*b*	*c*	*d*
1.	$4 = 3\frac{12}{12}$ $-\ 3\frac{7}{12} = 3\frac{7}{12}$ $\frac{5}{12}$	7 $-\ 1\frac{5}{8}$	34 $-\ 12\frac{4}{18}$	23 $-\ 15\frac{3}{11}$
2.	$8\frac{1}{4} = 7\frac{5}{4}$ $-\ 6\frac{3}{4} = 6\frac{3}{4}$ $1\frac{2}{4} = 1\frac{1}{2}$	$9\frac{5}{6}$ $-\ 7\frac{1}{6}$	$3\frac{9}{10}$ $-\ 1\frac{7}{10}$	$6\frac{2}{7}$ $-\ 3\frac{4}{7}$
3.	$23\frac{2}{9} = 23\frac{4}{18}$ $-\ 16\frac{1}{18} = 16\frac{1}{18}$ $7\frac{3}{18} = 7\frac{1}{6}$	$3\frac{5}{6}$ $-\ 2\frac{2}{3}$	$27\frac{3}{8}$ $-\ 15\frac{1}{4}$	$7\frac{5}{6}$ $-\ 4\frac{1}{12}$
4.	$10\frac{5}{7} = 10\frac{15}{21}$ $-\ 4\frac{2}{3} = 4\frac{14}{21}$ $6\frac{1}{21}$	$24\frac{2}{6}$ $-\ 10\frac{4}{5}$	$17\frac{5}{6}$ $-\ 12\frac{2}{3}$	$36\frac{4}{5}$ $-\ 32\frac{1}{4}$
5.	$16\frac{1}{6} = 16\frac{4}{24} = 15\frac{28}{24}$ $-\ 9\frac{3}{8} = 9\frac{9}{24} = 9\frac{9}{24}$ $6\frac{19}{24}$	$7\frac{5}{9}$ $-\ 4\frac{1}{6}$	$8\frac{2}{5}$ $-\ 5\frac{2}{3}$	$14\frac{1}{4}$ $-\ 11\frac{3}{5}$

Line up the digits. Subtract. Simplify.

	a	*b*	*c*
6.	$13\frac{4}{9} - 7\frac{1}{2} =$ ________ $13\frac{4}{9}$ $-\ 7\frac{1}{2}$	$9\frac{7}{8} - 4\frac{7}{8} =$ ________	$24\frac{2}{5} - 16\frac{4}{5} =$ ________

PROBLEM SOLVING

Applications

Solve.

1. Renee used $1\frac{1}{2}$ dozen eggs to make deviled eggs. Tony pickled $1\frac{3}{4}$ dozen eggs. How many dozen eggs did they use in all?

Answer ______________

2. The Shins went on a trip. They traveled $\frac{1}{12}$ of the trip on the first day, and $\frac{5}{12}$ on the second day. How much more do they have to travel?

Answer ______________

3. One recipe calls for $3\frac{3}{4}$ cups of flour. Another recipe uses $2\frac{1}{3}$ cups of flour. What is the difference in the amounts of flour?

Answer ______________

4. Quinn bought $3\frac{3}{4}$ pounds of cheese and $5\frac{5}{8}$ pounds of fish at the store. How many pounds of food did he buy in all?

Answer ______________

5. Sam is $5\frac{5}{12}$ feet tall. How much would he have to grow to be 6 feet tall?

Answer ______________

6. Enrico mailed 2 packages. One weighed $4\frac{2}{5}$ pounds and the other weighed $5\frac{4}{5}$ pounds. How much did the packages weigh in all?

Answer ______________

7. Sable spent $3\frac{1}{4}$ hours reading a book, $1\frac{7}{8}$ hours playing the piano, and 1 hour playing a game. How many hours did she spend doing these activities?

Answer ______________

8. Raj walks $\frac{3}{8}$ mile to school. Kevin walks $\frac{1}{4}$ mile to get to school. Who walks the furthest distance? By how much?

Answer ______________

PROBLEM-SOLVING STRATEGY

Make a Table

For some problems, it is necessary to make many calculations before finding the final answer. It is easier to solve this type of problem if the information is organized in a table.

Read the problem.

An art supply store sells color pencils for 40¢ each and erasers for 70¢ each. How many of each can you buy for exactly $3.40?

Make a table.

Calculate the price of different combinations of color pencils and erasers. Organize the results in a table. For example, here is how to find the cost of 3 color pencils and 2 erasers.

$$(3 \times \$0.40) + (2 \times \$0.70) = \$1.20 + \$1.40 = \$2.60$$

Number of Erasers \ Number of Color Pencils	0	1	2	3	4	5
0	$0	$0.40	$0.80	$1.20	$1.60	$2.00
1	$0.70	$1.10	$1.50			
2	$1.40			$2.60	$3.00	$3.40
3	$2.10			$3.30	$3.70	
4	$2.80	$3.20	$3.60			
5	$3.50					

Solve the problem.

The table shows that you can get 5 color pencils and 2 erasers for exactly $3.40.

Solve.

1. Complete the table shown above.

2. Eric spent exactly $2.50 for color pencils and erasers at the art supply store. Use the table to find out the number of erasers that he bought. (Hint: Use the table above.)

Answer ______________________

3. Lauren spent exactly $4.10 for color pencils and erasers at the art supply store. Use the table to find out the number of color pencils that she bought. (Hint: Use the table above.)

Answer ______________________

Solve.

4. A store sells containers of fruit yogurt for $1.19 and plain yogurt for $0.99. Complete the table at the right to show the combinations.

Plain Yogurt (columns) / Fruit Yogurt (rows)

	0	1	2	3
0	$0	$0.99		
1	$1.19			
2				
3				

5. Mitch spent $6.54. How many of each type did he buy? (Hint: Use your completed table.)

Answer ______________

6. Lisa spent $3.37. How many of each type did she buy? (Hint: Use your completed table.)

Answer ______________

7. Carl emptied 25 pennies and dimes from his pockets. He had $1.33 in all. How many pennies did he have in his pockets?

Dimes	Pennies	Total Value	
22	3	$2.23	No

Answer ______________

8. Gretchen has 10 coins in her purse with a total value of $1.60. The coins are dimes and quarters. How many of each coin does she have?

Quarters	Dimes	Total Value	
1	9	$1.15	No

Answer ______________

9. Minnie can spend exactly $1.00 for some of these items.

Rulers, 10¢ each

Pencils, 20¢ each

Folders, 50¢ each

What can she buy? Use the table to show all the possibilities.

Folders 50¢	Pencils 20¢	Rulers 10¢	Total

REVIEWING BASIC SKILLS

Multiplication of Fractions

Multiply. Simplify.

	a	*b*	*c*	*d*
1.	$\frac{3}{5} \times \frac{2}{5} = \frac{6}{25}$	$\frac{5}{9} \times \frac{7}{9} =$	$\frac{3}{8} \times \frac{7}{8} =$	$\frac{6}{7} \times \frac{3}{7} =$
2.	$\frac{3}{7} \times \frac{8}{9} = \frac{8}{21}$ $\frac{\not{3}^1}{7} \times \frac{8}{\not{9}_3} = \frac{8}{21}$	$\frac{2}{5} \times \frac{7}{8} =$	$\frac{6}{7} \times \frac{2}{3} =$	$\frac{6}{7} \times \frac{3}{5} =$
3.	$8 \times \frac{2}{5} = 3\frac{1}{5}$ $\frac{8}{1} \times \frac{2}{5} = \frac{16}{5} = 3\frac{1}{5}$	$\frac{1}{9} \times 6 =$	$5 \times \frac{6}{7} =$	$\frac{3}{4} \times 2 =$
4.	$4\frac{2}{7} \times \frac{4}{5} = 3\frac{3}{7}$ $\frac{\not{30}^6}{7} \times \frac{4}{\not{5}_1} = \frac{24}{7} = 3\frac{3}{7}$	$\frac{7}{9} \times 3\frac{4}{5} =$	$6 \times 8\frac{5}{8} =$	$7\frac{1}{8} \times 8 =$
5.	$4\frac{4}{8} \times 8\frac{8}{9} = 40$ $\frac{\not{36}^4}{\not{8}_1} \times \frac{\not{80}^{10}}{\not{9}_1} = \frac{40}{1} = 40$	$3\frac{2}{5} \times 16\frac{2}{7} =$	$3\frac{1}{9} \times 7\frac{3}{4} =$	$9\frac{1}{16} \times 4\frac{1}{2} =$

Multiply. Simplify.

	a	*b*	*c*
6.	$\frac{8}{9} \times \frac{7}{8} \times \frac{4}{7} = \frac{4}{9}$ $\frac{\not{8}^1}{9} \times \frac{\not{7}^1}{\not{8}_1} \times \frac{4}{\not{7}_1} = \frac{4}{9}$	$\frac{5}{6} \times \frac{1}{9} \times \frac{9}{15} =$	$\frac{2}{5} \times \frac{1}{8} \times \frac{4}{7} =$

REVIEWING BASIC SKILLS

Division of Fractions

Remember, to divide by a fraction, multiply by its reciprocal.

Divide. Simplify.

	a	*b*	*c*	*d*
1.	$\frac{2}{5} \div \frac{3}{5} = \frac{2}{3}$ $\frac{2}{\cancel{5}_1} \times \frac{\cancel{5}^1}{3} = \frac{2}{3}$	$\frac{7}{8} \div \frac{7}{8} =$	$\frac{8}{9} \div \frac{4}{9} =$	$\frac{5}{6} \div \frac{1}{6} =$
2.	$\frac{3}{7} \div \frac{5}{9} = \frac{27}{35}$ $\frac{3}{7} \times \frac{9}{5} = \frac{27}{35}$	$\frac{5}{6} \div \frac{1}{8} =$	$\frac{2}{7} \div \frac{1}{3} =$	$\frac{3}{16} \div \frac{3}{8} =$
3.	$2 \div \frac{5}{6} = 2\frac{2}{5}$ $\frac{2}{1} \times \frac{6}{5} = \frac{12}{5} = 2\frac{2}{5}$	$\frac{8}{9} \div 5 =$	$3 \div \frac{3}{5} =$	$\frac{1}{2} \div 4 =$
4.	$4 \div 2\frac{7}{11} = 1\frac{15}{29}$ $\frac{4}{1} \times \frac{11}{29} = \frac{44}{29} = 1\frac{15}{29}$	$5\frac{3}{4} \div 3 =$	$7 \div 8\frac{1}{3} =$	$3\frac{2}{3} \div 8 =$
5.	$3\frac{3}{4} \div \frac{2}{5} = 9\frac{3}{8}$ $\frac{15}{4} \times \frac{5}{2} = \frac{75}{8} = 9\frac{3}{8}$	$\frac{3}{13} \div 4\frac{4}{26} =$	$5\frac{2}{5} \div \frac{4}{5} =$	$\frac{3}{8} \div 8\frac{7}{16} =$
6.	$3\frac{4}{9} \div 2\frac{1}{8} = 1\frac{95}{123}$ $\frac{31}{9} \times \frac{8}{17} = \frac{248}{123} = 1\frac{95}{123}$	$8\frac{1}{6} \div 5\frac{7}{8} =$	$4\frac{2}{5} \div 6\frac{3}{5} =$	$2\frac{5}{8} \div 8\frac{3}{4} =$

Meaning of Percent

The symbol % is read as *percent*. Percent means *per hundred* or *out of one hundred*. Therefore, 78% means 78 per hundred or 78 out of 100, or 78 hundredths.

By writing the percent as hundredths, we can write a percent as a fraction or as a decimal.

To change a *percent to a decimal*, move the decimal point 2 places to the left and drop the percent sign (%). Write zeros as needed.

EXAMPLES

78% = 0.78　　001% = 0.01　　200% = 2.00

To change a *percent to a fraction*, place the percent over 100 and drop the % sign. Simplify.

EXAMPLES

$78\% = \frac{78}{100} = \frac{39}{50}$　　$1\% = \frac{1}{100}$　　$200\% = \frac{200}{100} = 2$

PRACTICE

Change each percent to a decimal and then to a fraction. Simplify.

	a	*b*
1.	32% = ______	5% = ______
2.	4% = ______	95% = ______
3.	21% = ______	60% = ______
4.	1% = ______	300% = ______
5.	100% = ______	7% = ______
6.	15% = ______	3% = ______
7.	89% = ______	75% = ______
8.	90% = ______	42% = ______

Changing Decimals and Fractions to Percents

To change a *decimal to a percent,* move the decimal point 2 places to the right and write a percent symbol. Write zeros as needed.

EXAMPLES

0.264 = *26.4%* 0.05 = *5%* 0.80 = *80%*

PRACTICE

Change the following to percents.

	a	*b*	*c*
1.	0.19 = *19%*	0.30 = ______	0.65 = ______
2.	0.01 = ______	0.1 = ______	0.90 = ______
3.	0.75 = ______	0.18 = ______	0.09 = ______
4.	0.05 = ______	0.183 = ______	0.2 = ______
5.	0.10 = ______	0.6 = ______	0.41 = ______

To change a *fraction to a percent,* first change the fraction to a decimal by dividing the numerator by the denominator. Then rewrite the decimal quotient as a percent.

Write $\frac{4}{5}$ as a percent.

Divide.

$5\overline{)4.00}$ = 0.80 = *80%*

Write $\frac{1}{8}$ as a percent.

Divide.

$8\overline{)1.000}$ = 0.125 = *12.5%* = *$12\frac{1}{2}\%$*

PRACTICE

Change the following to percents.

	a	*b*	*c*
6.	$\frac{1}{5}$ = *20%*	$\frac{9}{10}$ = ______	$\frac{1}{2}$ = ______
7.	$\frac{7}{10}$ = ______	$\frac{3}{4}$ = ______	$\frac{1}{4}$ = ______
8.	$\frac{1}{16}$ = ______	$\frac{3}{8}$ = ______	$\frac{2}{16}$ = ______

Interchanging Fractions, Decimals, and Percents

Fill in the blanks below as illustrated in the first example.

-Work Space-

1. $\frac{1}{10}$ = *0.10* = *10%*

2. $\frac{1}{2}$ = ________ = ________

3. ________ = 0.75 = ________

4. ________ = 0.30 = ________

5. ________ = ________ = 45%

6. ________ = ________ = 20%

Solve.

7. Computer sales are up $26\frac{1}{2}\%$. Write $26\frac{1}{2}\%$ as a decimal.

Answer ________

8. Thomas spends twenty percent of his salary for food. What fraction of his income does he spend on food?

Answer ________

9. Janet told her class that 18 of the 20 varieties of flowers she planted have bloomed. What percent of Janet's flowers have bloomed?

Answer ________

10. Some city and state sales taxes total 8%. Write the decimal a salesclerk would use to find the tax on a purchase.

Answer ________

11. Fifty percent of the Record Club members attended the last meeting. Write fifty percent as a fraction.

Answer ________

12. In the U.S., two out of five families have more than one car. Write this fraction as a decimal.

Answer ________

Percents Greater Than 100% and Less Than 1%

Large percents are used just like the small percents we have been using. Proceed as you did with small percents. To change the percent to a decimal, move the decimal point 2 places to the left. Drop the percent sign.

EXAMPLES

100% = 1.00 = 1 225% = 2.25 $142\frac{1}{2}\%$ = 142.5% = 1.425

PRACTICE

Change each percent to a decimal.

	a	*b*	*c*
1.	600% = 6	$250\frac{1}{2}\%$ = ______	275% = ______
2.	225% = ______	125% = ______	$187\frac{1}{2}$ = ______
3.	101% = ______	300% = ______	199% = ______

To solve problems using percents smaller than 1%, change the fraction to a decimal.

EXAMPLES

$1\% = \frac{1}{100} = 100\overline{)1.00}$ → 0.01

$\frac{1}{2}\% = 0.5\% = \frac{0.5}{100} = 100\overline{)0.500}$ → 0.005

PRACTICE

Change each percent to a decimal.

	a	*b*	*c*
4.	$\frac{1}{4}\%$ = 0.0025	$\frac{2}{5}\%$ = ______	$\frac{1}{10}\%$ = ______
5.	$\frac{5}{16}\%$ = ______	$\frac{5}{8}\%$ = ______	$\frac{4}{5}\%$ = ______
6.	$\frac{1}{8}\%$ = ______	$\frac{3}{4}\%$ = ______	$\frac{3}{10}\%$ = ______

Finding a Percent of a Number

To find a percent of a number, write a percent sentence. Every percent sentence consists of three numbers: the rate, the whole, and the part.

15% of 80 = 12

rate (15%), whole (80), part (12)

If the ***part*** is missing in a percent problem, solve by first changing the *rate* to a decimal. Then multiply the rate by the *whole*.

Remember, "of" means multiply.

Find: 38% of 70

$38\% \times 70 = ?$
$0.38 \times 70 = 26.6$

Find: 4% of 57

$4\% \times 57 = ?$
$0.04 \times 57 = 2.28$

PRACTICE

Change each percent to a decimal. Solve.

	a	*b*
1.	15% of 60 $15\% \times 60 =$ $0.15 \times 60 = 9$	5% of 400
2.	28% of 120	20% of 60
3.	14% of 500	6% of 150
4.	8% of 80	35% of 200

5. Yolanda gets a 15% discount on purchases at the store where she works. What will her discount be on a purchase of $37.80?

Answer ____________

6. State and city sales tax amounts to 8%. How much tax will Marty pay on the purchase of a watch costing $58.50?

Answer ____________

Finding a Percent of a Number

Another way to find the missing ***part*** in a percent problem is to use fractions rather than decimals. Change the *rate* to a fraction. Multiply the fraction by the *whole*.

Find: 20% of 35

$$20\% = 0.20 = \frac{1}{5}$$

$$\frac{1}{5} \times 35 = 7$$

Find: 75% of 16

$$75\% = 0.75 = \frac{3}{4}$$

$$\frac{3}{4} \times 16 = 12$$

PRACTICE

Change each percent to a fraction. Solve.

	a	*b*
1.	50% of 60	25% of 400
2.	75% of 120	$33\frac{1}{3}\%$ of 45
3.	10% of 180	20% of 60
4.	40% of 150	60% of 160

5. During a 40%-off sale, Tina bought a dress which normally sold for $42.60. How much did she save by buying the dress while it was on sale?

Answer ____________________

6. Leonard set aside 20% of his salary for insurance and savings. If he earned $16,000, how much did he set aside?

Answer ____________________

Finding a Number When a Percent of It Is Known

To find the ***whole*** in a percent problem, write a percent sentence. Change the *rate* to a decimal. Divide the *part* by the decimal.

35% of what number is 70?

35% of ? = 70
? = 70 ÷ 0.35
200 = 70 ÷ 0.35

25% of what number is 60?

25% of ? = 60
? = 60 ÷ 0.25
240 = 60 ÷ 0.25

PRACTICE

Change each percent to a decimal. Solve.

	a	*b*
1.	25% of what number is 10? 25% of ? = 10 ? = 10 ÷ 0.25 40 = 10 ÷ 0.25	40% of what number is 100?
2.	10% of what number is 1.7?	200% of what number is 10?
3.	150% of what number is 150?	20% of what number is 60?
4.	25% of what number is 20?	2% of what number is 40?
5.	60% of what number is 9?	80% of what number is 4?
6.	25% of what number is 40?	100% of what number is 15?

Finding What Percent One Number Is of Another

To find the ***rate*** in a percent problem, write a percent sentence. Divide the *part* by the *whole*. Then write the decimal answer as a percent.

What percent of 30 is 24?

$?\% \times 30 = 24$
$? = 24 \div 30$
$0.8 = 24 \div 30$
$0.8 = 80\%$

What percent of 60 is 15?

$?\% \times 60 = 15$
$? = 15 \div 60$
$0.25 = 15 \div 60$
$0.25 = 25\%$

PRACTICE

Find the rate.

a

1. What percent of 64 is 16?

$?\% \times 64 = 16$
$? = 16 \div 64$
$0.25 = 16 \div 64$
$0.25 = 25\%$

2. What percent of 700 is 14?
3. What percent of 90 is 30?
4. What percent of 500 is 5?
5. What percent of 6000 is 600?
6. What percent of 184 is 92?

b

1. What percent of 28 is 14?
2. What percent of 100 is 9?
3. What percent of 85 is 17?
4. What percent of 202 is 101?
5. What percent of 500 is 50?
6. What percent of 100 is 1?

Percent of Decrease

To find the current value or amount, first multiply the original amount by the percent of decrease. Then subtract.

EXAMPLE Bill bought a car last year for $11,000. When he tried to sell it this year, he found the value of the car had decreased 20%. What is the current value of the car?

original value	→	*$11,000*	original value →	*$11,000*
percent of decrease	→	× *0.20*	amount of decrease →	− *2200*
amount of decrease	→	*$2200*	current value →	*$8800*

Solve.

1. A storekeeper decreased the price of a coat by 20%. If the coat had been $85, what was the sale price?

Answer ____________________

2. 660 people attended the opening basketball game. The attendance dropped 25% for the second game. How many people attended the second game?

Answer ____________________

3. Last month the Brown family's electric bill was $68. This month their bill decreased by 12%. How much was their bill?

Answer ____________________

4. Last year, the track team had 48 members. This year the membership dropped by 25%. How many people are on the track team this year?

Answer ____________________

5. Last year, Jenny hit 15 home runs. This year her number of home runs decreased by 20%. How many home runs did Jenny hit this year?

Answer ____________________

6. Joseph's TV shop reduced the price of their best-selling television by 30%. The television normally sells for $800. What is the new price of the television?

Answer ____________________

Percent of Increase

To find the current value or amount, first multiply the original amount by the percent of increase. Then add.

EXAMPLE Terri raised 3200 bushels of oats last year. Because of better weather conditions this year, she expects to increase her crop by 25%. How many bushels does she expect this year?

original value →	3200 bu	original amount →	3200 bu
percent of increase →	×0.25	amount of increase →	+ 800 bu
amount of increase →	800 bu	amount expected →	4000 bu

Solve.

1. Last year a certain vacant lot cost $80,000. This year it will cost 10% more. How much will the lot cost this year?

   Answer ______________________

2. Laura earns $56.40 a day. If she gets a 10% raise, how much will she be earning?

   Answer ______________________

3. The number of employees at Harbor Construction rose by 16% between 1993 and 1994. In 1993, 25 people worked for Harbor Construction. How many people worked there in 1994?

   Answer ______________________

4. Last year, the Delaneys' monthly rent was $520. This year the rent was increased by 15%. What is the Delaneys' rent now?

   Answer ______________________

5. Lyle could type 50 words per minute on his first typing test. He increased this number by 14% on the second test. How many words per minute did Lyle type on the second test?

   Answer ______________________

6. After a brisk walk, Brenda's pulse rate rose by 10%. Her pulse rate was 70 beats per minute before the walk. What was her pulse rate after the walk?

   Answer ______________________

PROBLEM-SOLVING STRATEGY

Write a Number Sentence

A number sentence shows how numbers are related to each other. If one of the numbers is not known, write n instead of a number. To solve the number sentence, you may need to find a missing addend or missing factor. Study these examples.

EXAMPLE 1

Read the problem.

Sam combined some wheat flour and $2\frac{1}{4}$ cups of rye flour for a recipe. He used a total of 6 cups of flour. How much wheat flour did he use?

Write a number sentence.

In the number sentence, n is the unknown amount of wheat flour.

Wheat flour		Rye flour		Total flour
n	+	$2\frac{1}{4}$	=	6

Solve the problem.

To find n, the missing addend, subtract $2\frac{1}{4}$ from 6.

$$n = 6 - 2\frac{1}{4} = 3\frac{3}{4}$$

Since $3\frac{3}{4} + 2\frac{1}{4} = 6$, the amount of wheat flour is $3\frac{3}{4}$ cups.

EXAMPLE 2

Read the problem.

Fifty-five percent of the people at a meeting are wearing sneakers. If 11 people are wearing sneakers, how many people are at the meeting?

Write a number sentence.

In the number sentence, n is the number of people at the meeting. Write 55% as the decimal 0.55.

Percent		People at meeting		Number wearing sneakers
0.55	×	n	=	11

Solve the problem.

To find the missing factor, divide 11 by 0.55.

$$n = 11 \div 0.55 = 20$$

Since $0.55 \times 20 = 11$, the number of people at the meeting was 20.

Write a number sentence. Then solve the problem.

1. Molly Martin got a raise of 73¢ per hour. She now earns $9.38 per hour. What was her hourly rate before her raise?

   Number Sentence ____________________

   Answer ____________________

2. Twenty percent of the students in a class are wearing blue pants. If 10 students are wearing blue pants, how many students are in the class?

   Number Sentence ____________________

   Answer ____________________

3. A recipe calls for $\frac{3}{4}$ cups of margarine. If the recipe is tripled, how much margarine will be needed?

   Number Sentence ____________________

   Answer ____________________

4. This week, Bernie plans to work 24 hours. This is $\frac{2}{3}$ as many hours as he worked last week. How many hours did he work last week?

   Number Sentence ____________________

   Answer ____________________

5. Gary has jogged $1\frac{5}{8}$ miles along a hiking path that has a total length of $2\frac{1}{4}$ miles. If he wants to jog the entire path, how much farther does he have to jog?

   Number Sentence ____________________

   Answer ____________________

6. For a project, Ellen cut a long dowel rod into five equal pieces, each $4\frac{1}{2}$ inches long. What was the length of the dowel rod?

   Number Sentence ____________________

   Answer ____________________

7. One-tenth of the students in Mr. Albert's class had perfect attendance. If 3 students had perfect attendance, how many students are in the class?

   Number Sentence ____________________

   Answer ____________________

8. The length of a picture frame is 1.5 times the width. If the length is 24 inches, what is the width?

   Number Sentence ____________________

   Answer ____________________

Calculator Applications

A calculator can be used to do problems with percents. Some calculators have a percent key, [%]. However, it is just as easy to find percents without a percent key.

EXAMPLE 1 Of the 22 students in Jackie's math class, 7 students have red hair. What percent of the students have red hair? Round to the nearest percent.

Write a percent sentence: What percent of 22 is 7?

$? \times 22 = 7$ Divide: 7 [÷] 22 [=] [0.3181818]

To round the display to the nearest percent, first copy the digits over to the hundredths place (0.31). Then increase the hundredths digit by 1 if the thousandths digit is 5 or greater. So, this display rounds to 0.32 or 32%. So, 32% of the students in the class have red hair.

EXAMPLE 2 Of the 23 students in Jackie's math class, 35% have black hair. How many students have black hair?

Write a percent sentence: 35% of 23 is what number?

$0.35 \times 23 = ?$ Multiply: 23 [×] 0.35 [=] [8.05]

Round 8.05 to a whole number. The number of students who have black hair is 8.

Round the number in each display to the nearest percent.

1. [0.7941176] *79%*
2. [0.3548387] ______
3. [0.0489514] ______
4. [0.2222222] ______

Solve. Round to the nearest whole number or percent.

5. In a class of 28 students, 5 were absent last Monday. What percent were absent?

Answer ______

6. In a school of 344 students, 326 students were present last Tuesday. What percent were present?

Answer ______

7. Mr. Ritter's class has 34 students. Seventeen percent of the students have blonde hair. How many have blonde hair?

Answer ______

8. Mrs. Norton's class has 29 students. If 41% of the students are wearing sweaters, how many are wearing sweaters?

Answer ______

Unit 2 Review

Change each measurement.

	a	*b*	*c*
1.	6 lb = ____ oz	52 in. = ____ ft ____ in.	15 pt = ____ c
2.	6 pt 3 c = ____ c	19 oz = ____ lb ____ oz	5 yd = ____ ft

Find each answer. Simplify.

	a	*b*	*c*	*d*
3.	$\frac{4}{5} - \frac{7}{10} =$ ____	$\frac{9}{10} \times \frac{1}{3} =$ ____	$\frac{2}{3} + \frac{7}{9} =$ ____	$\frac{4}{7} \div \frac{3}{14} =$ ____
4.	$2\frac{6}{7} \div 4\frac{3}{5} =$ ____	$8\frac{1}{8} - 6 =$ ____	$\frac{5}{6} \times \frac{3}{4} =$ ____	$4\frac{2}{3} + 5\frac{4}{9} =$ ____
5.	$\frac{7}{8} + \frac{3}{4} =$ ____	$2\frac{5}{6} \times 5\frac{6}{7} =$ ____	$9\frac{1}{4} - 3\frac{3}{8} =$ ____	$\frac{1}{4} \div \frac{3}{16} =$ ____

Find each number.

	a	*b*
6.	Find 8% of 96.	What is 15% of 350?
7.	What percent of 32 is 0.8?	What percent of 80 is 24?
8.	What is 56% of 240?	5% of what number is 65?
9.	9% of what number is 45?	Find 42% of 98.

Hourly Wages

An **employee** is any person who earns money, or **wages,** to do work for another person or company. The **employer** is the person or company who pays the employee.

Some employees are paid an hourly wage. To find the total wages for a week, find the number of hours worked and multiply by the hourly wage.

EXAMPLE Carol is a cashier at Garden Records. She earns $6.50 per hour. She works from 1:00 P.M.-7:00 P.M. Tuesdays through Saturdays. How much does Carol earn per week?

Hours per day		Days per week		Hours per week
6	×	5	=	30
Total hours		**Hourly wage**		**Pay per week**
30	×	$6.50	=	$195.00

Carol earns $195.00 per week.

Find each person's weekly earnings. The hourly wage and the hours worked each day are listed in the chart.

	Name	Mon.	Tues.	Wed.	Thur.	Fri.	Sat.	Hours Worked	Hourly Wages	Weekly Earnings
1.	Louis Brandeis	8	8	8	8	8	0		$6.72	
2.	Elinor Woodward	7.5	7.5	4	7.5	7.5	4		$8.48	
3.	Norman Pucinski	0	8	8	5.5	8	8		$5.20	
4.	Thelma Watkins	6	7.5	7	8	6	0		$7.33	

5. James is a welder who earns $12.50 per hour. He works 8 hours per day, Monday through Friday. How much does James earn in one week?

Answer ______________

6. Anna is an installer for the telephone company. She earns $13.10 per hour. How much does she earn in 37.5 hours?

Answer ______________

7. Ted works in a paper-box factory. He receives $321.60 a week for 40 hours of work. Find his hourly wage.

Answer ______________

8. Yvonne's hourly wage is raised from $11.09 per hour to $12.35 per hour. She works 40 hours per week. How much more does she earn per week at the new rate?

Answer ______________

Hourly Wages: Overtime

Many employers pay a higher hourly wage for overtime work. Suppose an employee normally earns $6.30 per hour and works 40 hours per week. One week, the employee worked 48 hours and got overtime pay for the 8 extra hours, at 1.5 times the usual rate. How much will the employee earn for each hour of overtime?

$6.30 × 1.5 = $9.45 The overtime rate is $9.45 per hour.

Find the employees' total wages for the week.

Regular earnings	40 × $6.30 =	$252.00
Overtime earnings	8 × $9.45 =	$75.60
Total earnings		$327.60

The employees' total earnings for the week were $327.60.

Suppose that overtime pays time-and-a-half. That is, the hourly wage for overtime is 1.5 times the regular hourly wage. Find the hourly wage for overtime, given the regular hourly wage.

1. $8.20 **2.** $5.86 **3.** $12.66 **4.** $7.65

Complete the chart to find the total weekly earnings for each employee. The overtime rate is time-and-a-half. Round each amount to the nearest cent.

	Job	Hourly Wage	Regular Hours	Regular Earnings	Overtime Hours	Overtime Rate	Overtime Earnings	Total Earnings
5.	Printer	$10.30	40		6			
6.	Food server	$5.50	40		4.5			
7.	Carpenter	$14.88	40		10			
8.	Retail clerk	$8.95	40		2.5			

9. Rita Higgins is a nurse. Her regular rate is $14.62 per hour. She is paid double-time (two times the hourly rate) for holidays. During one week, she worked 36 regular hours plus 8 holiday hours. Find her total earnings for the week.

Answer ____________

Salary

Some employees earn a **salary,** or fixed amount, during each pay period. Suppose an employee's salary is $19,700 per year. What is the weekly salary? Divide by the number of weeks in a year.

Salary per year		Weeks per year		Salary per week
$19,700	÷	52	=	$378.85

Complete the chart. Round amounts to the nearest cent. (Hint: Find the monthly salary by dividing the yearly salary by 12.)

	Job	Weekly Salary	Yearly Salary	Monthly Salary
1.	Store Manager	$653.95		
2.	Librarian		$38,420.20	
3.	Dietician	$454.46		
4.	Biologist		$25,200.00	
5.	Drafter	$486.80		

6. Linda earns $2538 per month. Billie earns $32,700 per year. Who earns more, Linda or Billie?

Answer ____________

7. Phyllis earns a salary of $1830 per month. If she gets a 7% raise, how much more will she earn each month?

Answer ____________

8. Karl's salary is $30,500 per year. If he gets a 10% raise, what will his yearly salary be?

Answer ____________

9. Rodney's salary was raised $80 per week. What was his yearly increase?

Answer ____________

10. Ted sets aside 8% of his monthly salary of $2800 for retirement. Mary saves $50 a week from her salary. Who saves more in one year?

Answer ____________

11. An employee earns $1243 per month. What is the weekly salary? (Hint: Find the yearly salary before you find the weekly salary.)

Answer ____________

Wages and Tips

Workers in many service jobs earn tips from customers. Most waiters, taxi drivers, hair stylists, delivery drivers, and bellhops earn tips in addition to their hourly wage, which may be very low. By law, employees that earn tips regularly in their job must report the amount of tips to the employer. This is necessary because employees are required to pay taxes on this income as well as regular wages.

EXAMPLE Ruben Altaha is a waiter in a restaurant. He works about 7 hours each day at $2.10 per hour. On an average day he earns about $45 in tips. About how much does he earn in wages and tips each day? About how much does he earn per week if he works 5 days?

Wages		Tips		Total per day
($2.10 × 7)	+	$45	=	$59.70

Total per day		Days per week		Weekly earnings
$60	×	5	=	$300

Including tips, Ruben earns about $60 per day or $300 per week.

Solve.

1. A hair stylist at Pierre's hair salon earns $6.80 per hour plus tips. The stylist works 40 hours per week with tips of about $240 per week. Find the total weekly earnings.

Answer ______________________

2. Jerry is a parking lot attendant. He earns $2.50 per hour plus tips of about $8 per hour. About how much does he earn in a 40-hour week?

Answer ______________________

3. Helen is a waitress in Mama Lucia's Cafe. She earns $32 per day in wages, and her tips average $3 per table. Helen serves about 20 tables each day. Including tips, about how much does she earn per day?

Answer ______________________

4. Stella works as a manicurist. She works five 8-hour days per week, earning $4.50 per hour plus tips. On the average, she gets $15 per day in tips. Including tips, about how much does she earn each week?

Answer ______________________

5. Angelo delivers pizzas. He is paid $7.20 per hour plus tips. Here are the amounts earned from tips one week: Monday $22, Tuesday $30, Wednesday $27, Friday $23, Saturday $32. Find his total earnings for that week if he worked 40 hours.

Answer ______________________

6. Tahshi works part-time as a bellhop at a hotel. She earns $3.75 per hour and works 20 hours per week. Here are the amounts of her tips one week: $9, $8.50, $12.25, $15.75. Including tips, about how much does she earn per week?

Answer ______________________

Commission

Commission is a fee on a percentage basis for work done. Many persons, especially salespersons, work on commission alone or on a salary plus commission. If they work on commission alone, they are paid a certain **commission rate,** or percent of their sales.

EXAMPLE Kim sells houses for a real estate company. She earns $3\frac{1}{2}$% of the selling price of the property. If Kim sells a house for $85,000, how much is her commission? Multiply $85,000 by $3\frac{1}{2}$% or 3.5% or 0.035.

Commission rate		Selling price		Amount of commission
0.035	×	$85,000	=	$2975

Write each commission rate as a decimal. Then multiply to find the amount of commission.

	Amount of Sales	Commission Rate	Decimal	Amount of Commission
1.	$125,000	3%		
2.	$96,400	2%		
3.	$100,000	2.5%		
4.	$68,900	3.5%		
5.	$174,100	$1\frac{1}{2}$%		
6.	$32,491	5%		

7. Serene sells 1-year magazine subscriptions. She earns a commission of 25% of her sales. One evening, she sold $236 worth of subscriptions. How much commission did she earn?

Answer ______________________

8. Clyde Edison sells encyclopedias. He earns a commission of 15% for each sale. How much does he earn if the amount of the sale is $799.50?

Answer ______________________

9. Lena Chiu sells cosmetics and receives commissions of 45%. She sold $95 worth of cosmetics to one customer. How much commission did she earn?

Answer ______________________

10. Bill Barnes sells cleaning products on commission. He earns 40% commission on his sales. If he has sales of $983, how much commission does he earn?

Answer ______________________

Salary plus Commission

Some salespersons earn both commission and a salary or hourly wage. The commission may be on all sales, or just sales above a certain amount.

EXAMPLE In a men's clothing store, Pete is paid $248 per week and $10\frac{1}{2}$% on all sales over $2500. One week he sold a total of $3760. How much did he earn in salary plus commission that week?

Sales for which commission is paid: $3760 − $2500 = $1260
Commission: 0.105 × $1260 = $132.30
Salary plus commission: $248 + $132.30 = $380.30

Solve.

1. Rita sells cars. She earns a salary of $800 per month, plus a commission of 2% on all sales. In December, Rita sold $149,000 worth of cars. What were her total earnings in December?

Answer ____________________

2. Elmer works in a computer store. He is paid a weekly salary of $180 and a commission of 4.5% of all sales. One week, he sold $9770 worth of merchandise. How much did Elmer earn that week?

Answer ____________________

3. Laura Lucas works in a furniture store. She earns $160 a week plus 6% of all sales over $2000. One week in January, Laura's total sales were $9400. What were Laura's earnings for that week?

Answer ____________________

4. Ted sells textbooks to schools. His company pays him a monthly salary of $980, plus a commission of $2\frac{1}{2}$% on the amount of sales. One busy month, Ted sold $213,000 worth of books. How much did he earn that month?

Answer ____________________

5. Carey has two job offers for sales jobs. For job A, he would earn a monthly salary of $2500 and no commission. For job B, he would earn $1000 per month and commission of 5% on all sales. Which would pay more if his sales are $50,000 per month?

Answer ____________________

6. Roberta works in a home appliance store. She is paid $195 per week and 3% on all sales over $1000. During one year, her average weekly sales were $7800. About how much did she earn each week? About how much did she earn during the year?

Answer ____________________

EARNING MONEY

Making Bar Graphs

Bar graphs are useful for comparing quantities. The table below shows last month's sales by representatives of Superior Office Equipment.

Name of Representative	October Sales
Andy Avalos	$15,300
Bonnie Banes	$31,800
Carl Casper	$35,200
Donna Dixon	$19,000
Edward Evans	$26,500
Faye Farmer	$24,100

To make a bar graph from the information in the table, draw a bar beside each person's name to indicate the amount of sales. The bar beside *A. Avalos* is shown. Notice that it extends past the mark for 14 (thousands) on the horizontal scale. It is closer to 16 than to 14.

OCTOBER SALES BY SALES REPRESENTATIVE (IN THOUSANDS)

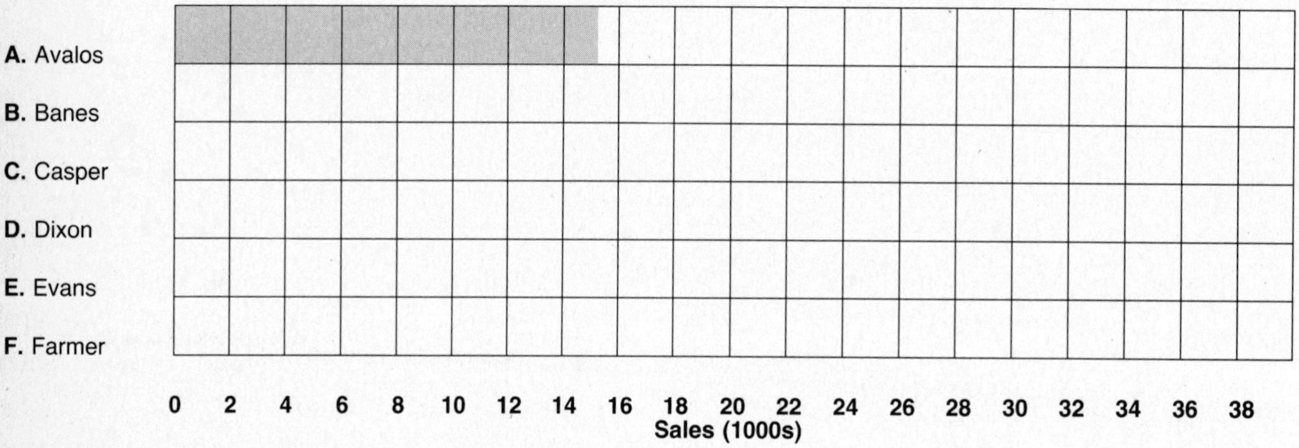

A. Avalos
B. Banes
C. Casper
D. Dixon
E. Evans
F. Farmer

0 2 4 6 8 10 12 14 16 18 20 22 24 26 28 30 32 34 36 38
Sales (1000s)

Answers

1. Complete the bar graph.
2. Each sales representative earns 15% commission on the total amount of sales. Which representative earned the most commission in October? ____________
3. How much commission was earned by the sales representative who sold the most? ____________
4. Which sales representative earned the least commission in October? ____________
5. How much commission was earned by the sales representative who sold the least? ____________
6. How many sales representatives had commissions of more than $3,000 in October? ____________

Employee Medical Insurance

Many employers offer group medical insurance to employees. This is an important benefit, since medical insurance provides financial protection in the event of a serious illness or injury. The company buys a group insurance policy and pays **premiums** for each employee to the insurance company. The insurance company then agrees to pay certain medical costs of the employees. Part or all of the cost of premiums may be deducted from an employee's pay.

The premiums for medical insurance depend on the specific costs that the insurance company will pay, the number of individuals covered by the insurance, and other factors. The table at the right shows premiums for two different medical insurance plans from *First Medical Insurance.*

—FIRST MEDICAL INSURANCE— Monthly Premiums		
	Plan A	Plan B
Single coverage	$125	$160
Family coverage	$205	$238

EXAMPLE Lila Shepard's employer pays 80% of the cost of group insurance through First Medical. Lila has Plan B, single coverage. How much is the employer's contribution each month for Lila's coverage? How much does Lila contribute each month?

Premium		Employer's percent		Employer's contribution
$160	×	0.80	=	$128

The employer pays $128. So, Lila pays $160 − $128 or $32.

Use the information above to complete the table.
(Hint: To find the annual contribution, multiply by 12.)

	Employee's Name	Plan	Monthly Premium	Employer's Percent	Employer's Contribution		Employee's Contribution	
					Monthly	Yearly	Monthly	Yearly
1.	Molly Rand	Family Plan A		50%				
2.	Brad Voltman	Single Plan A		90%				
3.	Chris West	Single Plan B		75%				
4.	Blake Horton	Family Plan B		20%				

5. Trish has two job offers. Company A offered her a salary of $19,000 and paid medical coverage worth $150 per month. Company B offered a salary of $1700 per month with no paid medical coverage. Which offer seems to be better?

Answer ____________________

6. Brent Goldman's employer pays 60% of the monthly medical premium of $234.00. The rest of the premium is deducted in equal portions from Brent's pay. If he is paid twice per month, how much is deducted each time?

Answer ____________________

Payroll Deductions

The amount an employee earns during a pay period is called the **gross pay.** However, the employer is required by the government to hold back certain amounts for federal taxes, state taxes, and social security (FICA). In addition, most employers hold back certain amounts for insurance. The employer pays these amounts to the government agencies and insurance companies on behalf of the employee.

The employee's paycheck is reduced by the amounts withheld. These payments for taxes and insurance are called **deductions.** The amount remaining after deductions are subtracted from earnings is the **net pay,** or take-home pay.

An employee's paycheck stub is shown below. What amount should be in the box for NET PAY?

EMPLOYEE NUMBER	PERIOD ENDING	HOURS	
		REGULAR	OVERTIME
108638	6/17/94	40	0

EARNINGS		
REGULAR	OVERTIME	GROSS PAY
364.40	0	364.40

DEDUCTIONS				NET PAY
FED. TAX	FICA	STATE TAX	MEDICAL INSURANCE	
41.86	27.37	10.93	15.50	

Total deductions: $41.86 + $27.37 + $10.93 + $15.50 = $95.66

Gross pay		Deductions		Net pay
$364.40	−	$95.66	=	$268.74

Complete each row of the chart to find the employee's net pay.

		Deductions					
	Gross Pay	**Fed. Tax**	**State Tax**	**FICA**	**Med. Ins.**	**Total Ded.**	**Net Pay**
1.	$550.00	$82.50	$22.00	$41.30	$7.50		
2.	$350.00	$32.00	$10.50	$26.29	$7.25		
3.	$350.00	$24.19	$9.80	$26.29	$42.70		
4.	$209.20	$16.74	$4.20	$15.71	0		
5.	$779.53	$155.90	$28.89	$58.54	$25.61		
6.	$658.44	$98.72	$19.76	$49.49	0		

7. In one pay period, Karen had $318.80 in regular earnings and $95.64 in overtime earnings. Her employer withheld $135.90 for taxes and insurance. What was Karen's take-home pay for that pay period?

Answer ______________

Calculator Applications

When you multiply or divide with a calculator, the display may show several digits to the right of the decimal point. If the display is an amount of money, you will usually need to round to the nearest cent, dime, or dollar. Think about how accurate the answer needs to be.

EXAMPLE Molly is having dinner in a restaurant. She wants to leave a tip of about 15% of the bill. If the dinner costs $12.19, how much tip should she leave?

12.19 [×] .15 [=] [1.8285]

Molly has several choices of how to round the number in the calculator display. For each choice shown below, one digit is underlined. Recall that to round the display, you should increase the underlined digit by 1 if the digit to the right is 5 or greater. Digits to the right should be dropped or changed to zero.

Nearest cent	Nearest dime	Nearest dollar
1.8285	1.8285	1.8285
$1.83	**$1.80**	**$2**

Molly decides to leave a tip of $1.80.

Look at each calculator display. Round to the nearest cent, dime, and dollar.

		Nearest cent	Nearest dime	Nearest dollar
1.	1.0935	*$1.09*	*$1.10*	*$1*
2.	14.2575			
3.	4.8915			
4.	0.807			
5.	2.6805			
6.	3.668			
7.	0.3975			
8.	15.558			

9. Joe and three of his friends had dinner at a restaurant. If the bill was $55.48, how much was the 15% tip? Round to the nearest dime.

Answer ______________

10. Suppose Joe and his three friends evenly split the total cost of the dinner and tip. How much did each person contribute? Round to the nearest dollar.

Answer ______________

PROBLEM-SOLVING STRATEGY

Identify Substeps

Solving a problem requires several steps such as reading the problem, deciding what to do, and finding the answer. Some steps of solving the problem may need to be broken down into *substeps,* or smaller steps. Identifying the substeps is important in planning how to solve the main problem.

Read the problem.

In one week Wendy worked 42.5 hours. She earned overtime for each hour over 40. Her overtime rate was 1.5 times her regular rate of $8.70 per hour. What was her gross pay for the week?

Identify substeps.

1. How many hours of overtime pay did Wendy earn?
2. How much was she paid for each hour of overtime?
3. How much was the total overtime pay?
4. How much was the total pay for the 40 regular hours?

Solve the problem.

First find the answer for each substep.

1. $42.5 - 40 = 2.5$ hours overtime
2. $1.5 \times \$8.70 = \$13.05$ per hour of overtime
3. $2.5 \times \$13.05 = \$32.625$ or $32.63 total overtime pay
4. $40 \times \$8.70 = \$348.00$ total regular pay

Then solve the main problem.

Regular pay		Overtime pay		Gross pay
$348.00	+	$32.63	=	$380.63

Wendy's gross pay for the week was $380.63.

Write two substeps for each problem. Do not solve.

1. Georgette's gross pay for a week was $304.00. The deduction for state tax was 3% of her gross pay. Her other deductions totalled $56.24. What was Georgette's take-home pay for the week?

Substep 1 ______________________

____

Substep 2 ______________________

____

2. One week Billy worked 36 hours at $6.70 per hour. The next week he worked 40 hours at $6.85 per hour. How much more did he earn the second week?

Substep 1 ______________________

____

Substep 2 ______________________

____

Solve.

3. Solve each substep of Problem 1. Then solve the main problem.

Answer to Substep 1 ____________________

Answer to Substep 2 ____________________

Answer to main problem ____________________

4. Solve each substep of Problem 2. Then solve the main problem.

Answer to Substep 1 ____________________

Answer to Substep 2 ____________________

Answer to main problem ____________________

5. When Randy started his job, he earned $5.40 an hour. Now he makes $234 for working a 40-hour week. How much more does Randy earn per hour than he did when he started?

Answer ____________________

6. Evelyn earns a weekly salary of $560. Wanda earns a monthly salary of $2370. Who earns more per year? How much more?

Answer ____________________

7. Bonnie got a raise of 32¢ per hour. Now she earns $314.40 for a 40-hour week. What was her hourly rate before her raise?

Answer ____________________

8. Xavier earns a yearly salary of $16,500. If he gets a raise of $110 per month, what is the percent of increase?

Answer ____________________

9. One week, Christine earned 8% commission on sales of $4700. The next week, she earned 9% commission on sales of $5500. How much more commission did she earn the second week than the first week?

Answer ____________________

10. In one week, Quentin earned $72.90 for overtime hours. He earns time-and-a-half for overtime. If his regular rate is $8.10 per hour, how many hours of overtime did Quentin work?

Answer ____________________

Retail Store Profits

Another very common way of earning money is to operate a retail store. In order to make money, the store operator must set prices high enough to pay the cost of the merchandise and the expenses of operating the store. The **gross income** is the total amount of money taken in from sales of merchandise. After cost of merchandise and other expenses are subtracted, the amount left is the **profit.**

EXAMPLE Last year, the Green Front Grocery sold $250,000 worth of groceries. Cost of the groceries was $190,000. Rent, wages, utilities, and other expenses amounted to $40,000. What was the profit?

Gross income		Expenses		Profit
$250,000	−	($190,000 + $40,000)	=	$20,000

What percent of gross income was the profit?

Profit		Gross income		Percent
$20,000	÷	$250,000	=	0.08 or 8%

Solve.

1. The Ideal Furniture Store did a total business of $156,000 during one month. The merchandise costs were $97,000. Other expenses amounted to $26,000. What was the profit?

Answer ____________

2. A shoe merchant did a total business of $39,500. Cost of merchandise was $19,300, and expenses were $11,800. What was the profit?

Answer ____________

3. The Breceda Meat Market did a total business of $52,000 in March. Cost of the meat was $35,000. Expenses amounted to $12,000. How much profit was made?

Answer ____________

4. In March, what percent of Breceda Meat Market's gross income was profit?

Answer ____________

5. Classic Clothing had total sales of $57,000 for April. Merchandise costs were $34,000. Other expenses were $11,000. What was the profit?

Answer ____________

6. In April, what percent of Classic Clothing's gross income was profit?

Answer ____________

Markup, Overhead, and Profit

The difference between the cost and selling price is called **markup.** Expenses incurred in selling, such as rent, heating, lighting, wages, etc., are called **overhead.** Thus, markup must include both overhead and profit, as shown in this diagram.

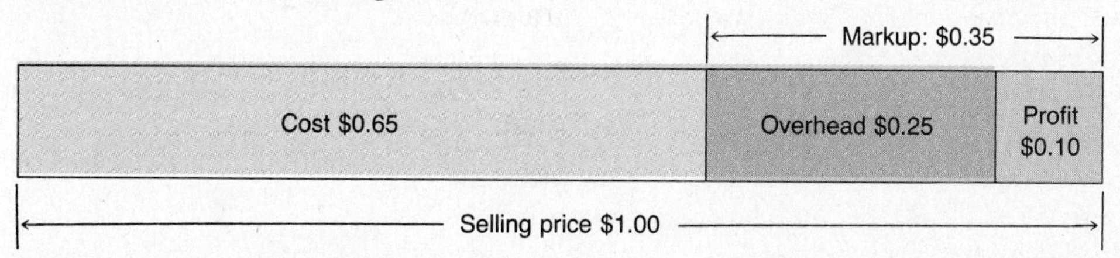

After years of experience, a store owner or manager can estimate how much the overhead will be and how much markup to allow between the cost and the selling price. Sometimes the markup is expressed as a percent of the cost of an item.

EXAMPLE Suppose a store owner pays $250 for a table, and allows a markup of 50% of the cost. For what price should the table be sold?

Markup = 50% of $250 = $125
Selling Cost = $250 + $125 = $375

The table should be sold for $375.

Solve.

1. A store manager paid $40 each for coats. The markup is 60%. For what price should the coats be sold?

Answer ____________

2. A store owner bought belts at $96 per dozen. If the markup is 45%, what price should be marked on each belt? (Hint: First find the cost for each belt.)

Answer ____________

3. Fresh foods must have a larger markup, since they are perishable. A grocer bought bananas for 32¢ per pound. If the markup is 100%, how much should the bananas be sold for?

Answer ____________

4. A music store bought a trumpet for $400 and used a markup of 50% to set the selling price. A customer bought the trumpet on sale for 50% off. How much did the customer pay?

Answer ____________

5. A store owner paid $35 each for sweaters. They were sold for $55. The overhead was $10. What was the profit on each sweater?

Answer ____________

6. A men's store bought summer suits at $120, to be sold at $180. The suits left at the end of the summer were put on sale for $150. If the overhead was $40 per suit, what was the loss on each suit that was sold for $150?

Answer ____________

EARNING MONEY

Making a Line Graph

The table shows last year's monthly sales and overhead at Capitol Clothing. Sales are given in thousands of dollars. That is, 12 means $12,000 dollars. This data can be shown on a line graph. Trends and changes can be easily detected from the graph.

	Winter			Spring			Summer			Fall		
	Jan	Feb	Mar	Apr	May	June	July	Aug	Sept	Oct	Nov	Dec
Sales ($1000)	62	38	57	90	76	68	74	95	92	85	98	124
Overhead	29	24	24	22	20	20	28	30	25	27	30	35

This line graph has two lines, a solid line for sales and a dashed line for overhead. To finish the line showing the sales amounts, first make a dot for each month across from the correct place on the vertical scale. Then connect the dots with a solid line.

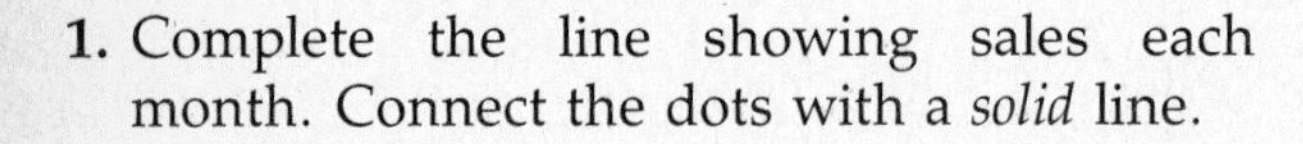

1. Complete the line showing sales each month. Connect the dots with a *solid* line.

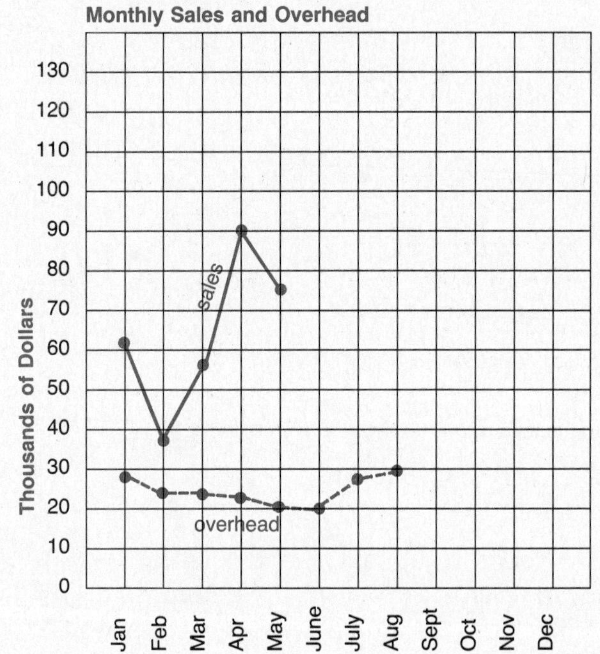

2. Complete the line showing overhead each month. Connect the dots with a *dashed* line.

3. Which month had the most sales? ____________________

4. Which month had the least sales? ____________________

5. Which month had the most overhead? ____________________

6. Which months had the least overhead? ____________________

7. Was more money spent on overhead during winter or spring? ____________________

8. Were there more sales during the summer months or during the fall months? ____________________

9. Between what two months did sales go up the most? ____________________

10. Between what two months did sales go down the most? ____________________

Self-Employment

Some people prefer to work for themselves rather than be an employee of another person or company. Here are some examples of people who are **self-employed.**

- Bob Cook has a mail-order business selling computer programs.
- Peter Seltzer does art work from his home.
- Dawn Fowler does childcare in her home.
- Toni Adams does lawncare for people in her neighborhood.
- Richard Riley sells reference books and earns commission from a company, but he plans his own work schedule and has an office in his own home.
- Mary Bennis gives music lessons from her home.

Self-employed persons are required to keep careful records and pay taxes on their profits. To determine the profits from self-employment, the expenses of doing business are deducted from earnings. Examples of expenses are costs for inventory (merchandise to sell), office equipment, phone service, mailing costs, supplies, and transportation. If a person operates a home-based business, a portion of the rent or mortgage may be counted as a business expense if certain conditions are met.

The chart shows last-year's self-employment expenses for the six people above. Find the total expenses for each person.

	Name	Inventory	Rent & Utilities	Equipment	Mail	Transportation	TOTAL EXPENSES
1.	B. Cook	$4780	$2422	$314	$549	$2409	$10,474
2.	P. Seltzer	0	1248	235	10	450	
3.	D. Fowler	0	3800	270	12	0	
4.	T. Adams	295	678	2504	58	4917	
5.	R. Riley	0	411	78	59	395	
6.	M. Bennis	0	0	15	10	35	

Suppose Bob Cook earned $20,400 during the year. To find his profit, subtract the expenses (from Exercise 1). His profit from self-employment was $20,400 – $10,474 or $9926.

For Exercises 7–11, copy each person's total expenses from the chart. Then subtract to find the profit.

	7. P. Seltzer	**8.** D. Fowler	**9.** T. Adams	**10.** R. Riley	**11.** M. Bennis
Earnings	$9700	$8450	$32,439	$4355	$2750
Expenses	–______	–______	–______	–______	–______
Profit	______	______	______	______	______

Profit Sharing

The Hi-Five Sports Equipment Company pays out 15% of its yearly profits as bonuses to employees with 5 or more years of service. These bonuses are distributed to employees through a **profit sharing** plan.

EXAMPLE Last year Hi-Five's profits were $1,642,000. How much profit was distributed to the employees?

$$\$1{,}642{,}000 \times 0.15 = \$246{,}300$$

The profit sharing plan is based on a point system, according to the number of years of service: 1 point for 5-9 years, 2 points for 10-14 years, and 3 points for 15 or more years. To find the amount of the bonus for each point, $246,300 was divided by the total number of points.

Complete the chart to find the total number of points.

	Years of Service	Number of Employees	Points Each	Multiply Number of Employees by Points Each
1.	5-9	41		
2.	10-14	25		
3.	15+	18		
			4.	TOTAL POINTS:

5. Divide $246,300 by the answer to Exercise 4 to find out how much each point is worth. Round to the nearest dollar.

Answer ____________________

6. Find the amount of bonus for an employee with 12 years of service. (Hint: The employee is awarded two points.)

Answer ____________________

7. Find the amount of bonus for an employee with 20 years of service.

Answer ____________________

8. Last year the ABX Company paid out 12% of its $3,377,000 profits in bonuses. Each of its 80 employees got an equal share. How much did each employee get?

Answer ____________________

9. The Janyln Company uses a point system for the profit-sharing plan. Of the 45 employees eligible for the plan in 1990, 13 received two points each and the rest received 1 point each. What was the total number of points?

Answer ____________________

10. The Janlyn Company paid out 10.5% of its profits in 1990. Their profits were $827,400. To the nearest cent, how much was each point worth? Use the total number of points from Exercise 9.

Answer ____________________

Business Travel Expenses

Mildred Jackson often travels on company business. She keeps a record of the money she spends and saves the receipts. She must also write down information such as the business purpose for the expenses and the locations where she stayed. When she returns, she turns in her records and is reimbursed, or paid back, by the company for the business expenses. The company must keep the records and receipts for tax purposes.

EXAMPLE On one business trip, Mildred stayed 3 nights at a hotel where the room rate was $65.50 per night, plus 8.9% hotel tax. To find the cost of her room for 3 nights, multiply. Then find the amount of tax and add.

Number of nights		Rate per night		
3	×	$65.50	=	$196.50

Hotel tax: 0.089 × $196.50 = $17.49
Total cost including tax: $196.50 + $17.49 = $213.99

Solve.

1. On another business trip, Mildred Jackson spent 4 nights at a hotel that cost $30.75 per night plus 11% hotel tax. What were her room expenses on this trip?

Answer ______________

2. Mildred Jackson's total cost, including tax, for hotel rooms on a 6-day trip was $459. What was the average cost per night?

Answer ______________

After business trips, Mildred Jackson is also paid back for her meals and transportation expenses. The tables at the right show meal and transportation expenses on a 3-day trip.

Complete the tables by finding the total of each row and each column. When you have finished, the right column and bottom row should have the same sum. Write this amount in the lower right corner.

MEALS

	Date	Breakfast	Lunch	Dinner	Totals
3.	*May 15*	*$4.75*	*$5.50*	*$9.50*	
4.	*May 16*	*$3.85*	*$6.00*	*$11.50*	
5.	*May 17*	*$4.50*	*$4.95*	*$10.50*	
6.	Totals				

TRANSPORTATION

	Date	Air Fares	Rental Cars & Gasoline	Totals
7.	*May 15*	*$315*	*$30*	
8.	*May 16*		*$30*	
9.	*May 17*	*$275*	*$30 + $5 gas*	
10.	Totals			

Disability Pay

Many companies provide **disability insurance** for their employees. If an employee becomes ill or injured and cannot work for an extended period of time, the insurance company pays part of the employee's salary. The illness or injury does not have to be job-related; it can be a result of a car accident or a medical problem.

Usually, the insurance company pays 50-80% of the disabled employee's salary each week for a defined period of time, such as 6 months. This **disability pay** ends after the defined period, or sooner if the employee is able to return to work.

EXAMPLE Shaun had surgery for a back injury and was unable to work for 30 weeks. His company's disability insurance pays 50% of an employee's weekly salary for up to 26 weeks. Shaun's weekly salary was $450. How much disability pay did he receive each week? What was the total amount of disability pay that he received?

Disability pay each week: $50\% \times \$450 = 0.5 \times \$450 = \$225$
Total disability pay (26 weeks): $26 \times \$225 = \$5850$

For each problem, find the employee's disability pay per week and the total disability pay.

	Disability pay per week	Total disability pay
1. Aretha is disabled for 25 weeks. Her weekly salary is $380. The disability insurance pays 50% of her salary for up to 1 year.	________	________
2. Corky is disabled for 36 weeks. His weekly salary is $360. The disability insurance pays 75% of his salary for up to 1 year.	________	________
3. George's disability insurance pays 80% of his salary for up to 18 months. He earns $950 per week, and is disabled for 42 weeks.	________	________
4. Kwan's disability insurance pays 50% of his salary for up to 2 years. His weekly salary is $578, and he is disabled for 81 weeks.	________	________
5. Paulette's disability insurance pays 60% of her salary for up to 24 months. Her weekly salary is $785. She is disabled for 65 weeks.	________	________
6. Donna has disability insurance that pays 80% of her weekly salary for up to 4 years. Her weekly salary is $1230. She is disabled for more than 5 years.	________	________

Retirement Income

Two sources of income for retired people are **social security** and **pensions.** More than 90% of American workers have a payroll deduction for social security tax. These workers and their spouses will receive monthly social security payments when they retire at age 62 or older. Social security payments alone are not usually enough to cover living expenses of a retired worker.

An employee who retires after working many years for the same company may get a pension from the company. The amount of a pension payment is based on the person's salary while employed, the age at retirement, and other factors.

In addition to social security and pensions, certain types of savings accounts and investments are designed to provide retirement income. By planning and saving money *before* retirement, a worker can make sure that retirement income will cover living expenses.

EXAMPLE Find the monthly income from a retirement account that provides annual income of $1860.

Yearly income		Months per yr		Monthly income
$1860	÷	12	=	$155

Solve.

1. Velma Parks gets $460 per month from social security, and has no other income. What is her yearly income?

Answer ____________

2. Sol Brandau gets a monthly pension of $984 and a monthly social security check of $475. What is his yearly income from these two sources?

Answer ____________

3. During last year, Jack Dooley received $3,290 from a retirement account. He also got $621 per month from social security. What was his average monthly income from these two sources?

Answer ____________

4. Before she retired, Victoria Hall's yearly salary was $18,200. Now she gets $605 per month from a pension plan and $525 from social security. How much less is she earning per year now?

Answer ____________

5. Brad Benson is 62 years old. If he waits until age 65 to retire, he will get $717 per month in social security payments. If he retires now, he will get 80% of that amount. How much will the payments be if Brad Benson retires now?

Answer ____________

6. Each month Marsha Mason gets $773 from her company pension and $532 from social security. Her living expenses are about $1550 per month. About how much money from savings will she need each month?

Answer ____________

PROBLEM-SOLVING STRATEGY

Use a Bar Graph

The graph shows the average weekly earnings of people in manufacturing industries. The information in the graph can be used to solve problems.

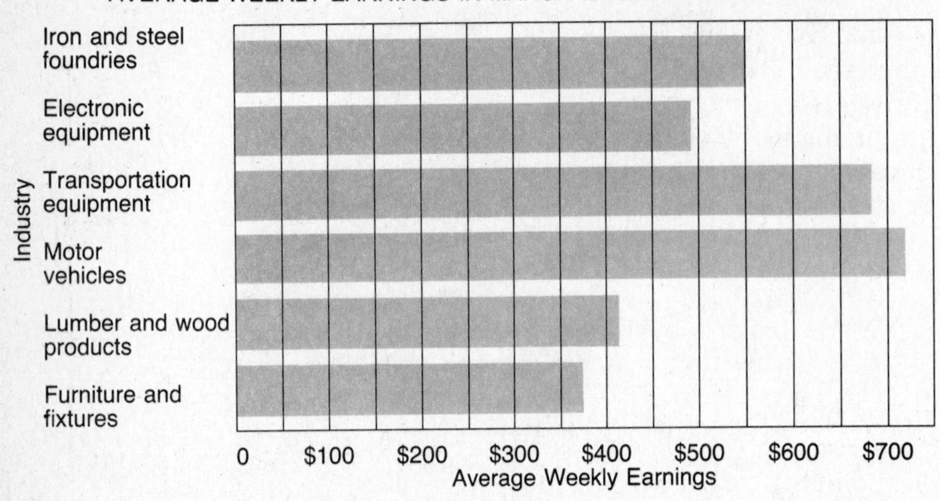

Read the problem.

Workers in the electronic equipment industry work about 41 hours per week. What is their average hourly wage?

Read the graph for data.

Look at the bar labeled electronic equipment and read the scale. If the end of the bar is between two numbers on the scale, estimate the length. Then write down the data.

Average weekly earnings: $490

Solve the problem.

Divide the weekly wage by the hours per week.

$490 ÷ 41 = $11.95

Solve. Use the bar graph above.

1. On the average, how much more is earned per week by workers in the transportation equipment industry than by workers in the lumber and wood products industry?

Answer ______________________

2. Workers in the furniture and fixtures industry work about 39 hours per week. What is their average hourly wage?

Answer ______________________

3. What is the average yearly salary for workers in iron and steel foundries?

Answer ______________________

4. What is the average yearly salary for workers in the motor vehicle industry?

Answer ______________________

PROBLEM-SOLVING STRATEGY

Use a Line Graph

A line graph may be used to show how data can change over a period of time. A double-line graph, such as the one below, is helpful for comparing sets of data.

Pat and Chris are hired by the same company. Each person will earn $10,000 the first year. Pat gets an increase of $2,000 each year. Chris gets an increase of 20% each year. The line graph at the right shows changes in their salaries over 5 years.

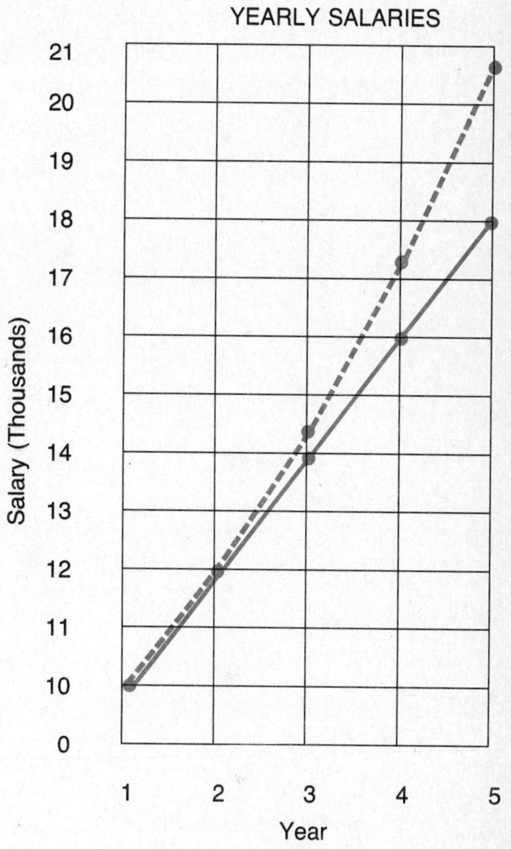

YEARLY SALARIES

Salary (Thousands)

0 10 11 12 13 14 15 16 17 18 19 20 21

Year

1 2 3 4 5

Read the problem.

In Year 4, about how much higher is Chris's salary than Pat's salary?

Read the graph for data.

Look above Year 4 on the horizontal scale. Find the point for Chris's salary. Trace an imaginary line from the point to the scale on the left. If the value is between two numbers on the scale, estimate. Write down the data for Chris's salary. Use the same method to find Pat's salary.

Chris's salary: about $17,300

Pat's salary: $16,000

Solve the problem.

$17,300 − $16,000 = $1,300

In Year 4, Chris earns about $1,300 more than Pat.

Solve. Use the line graph above.

1. In Year 5, about how much higher is Chris's salary than Pat's salary?

   Answer ______________

2. What is the percent of increase in Pat's salary from Year 3 to Year 4?

   Answer ______________

3. What is the percent of increase in Pat's salary from Year 4 to Year 5?

   Answer ______________

4. In Year 4, about how much is Chris's salary per week?

   Answer ______________

5. In Year 5, about how much is Pat's salary per week?

   Answer ______________

6. Predict the difference between Chris's salary and Pat's salary in Year 6.

   Answer ______________

Making Circle Graphs

A circle graph can help you compare parts of a whole, or quantities that total 100%. For example, a person's yearly income is a whole amount that can be represented by a circle. The circle can be divided into parts to represent way the income was earned. You would *not* make a circle graph from a list of incomes of several *different* people, because the people do not represent a whole.

If a circle is divided into 10 parts, each part is 10% of the whole. The circle graph at the right shows sources of Lola Dawson's income. The graph shows that 3 tenths or 30% was social security.

SOURCES OF LOLA DAWSON'S INCOME FOR 19—

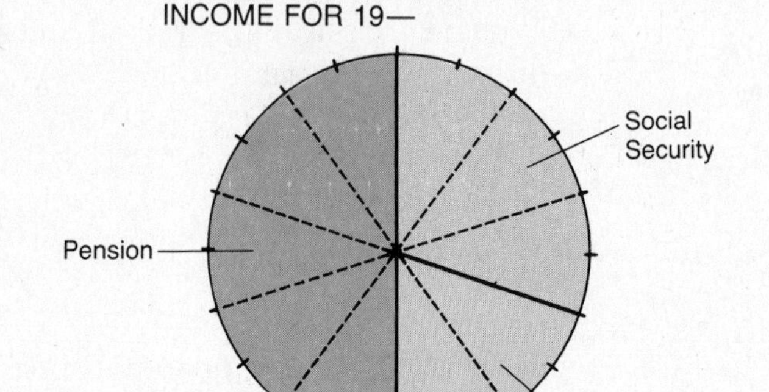

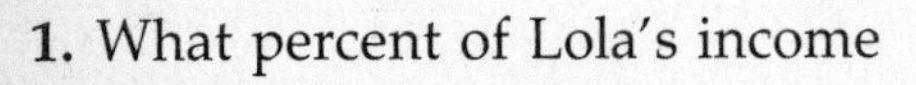

1. What percent of Lola's income was from her pension? ____________

2. What percent of Lola's income was from investments? ____________

To make a circle graph when you know the total amount and several parts, first find the percent of the total that each part represents.

EXAMPLE Randy Eisen's income from several sources for one year was \$15,900. Of the total, \$6960 was from social security. What percent of the total was from social security?

Social security		Total income		Decimal to nearest hundredth
\$6960	÷	\$15,900	=	0.438 or 0.44

Write the decimal as a percent. 0.44 = 44%

Randy Eisen's Income 19—	
Social security	\$ 6960
Pension	4134
Interest	1610
Part-time job	3196
TOTAL	\$15,900

Find the percent of Randy's income from each source. Use the table.

3. Pension ____________ **4.** Interest ____________ **5.** Part-time job ____________

6. Complete the circle graph of Randy Eisen's Income.

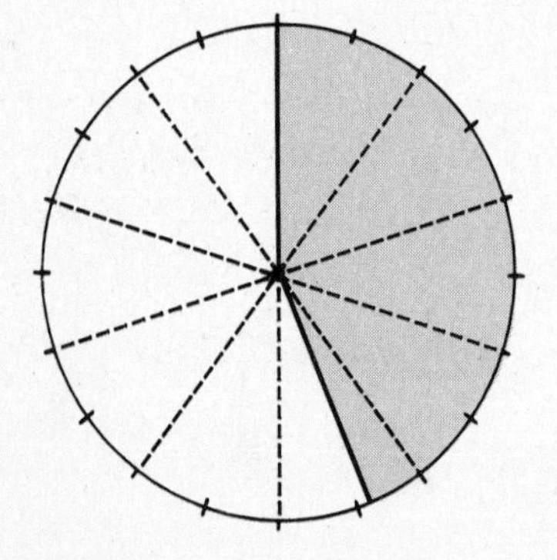

7. Did social security contribute more or less than half of Randy Eisen's income? ________

8. Which source contributed about one-fourth of Randy Eisen's income? ________

9. Which source contributed about one-fifth of the income? ________

Unit 3 Review

Solve.

1. Liz earns $432 for a 40-hour work week. She earns $12.48 for each hour of overtime. How much does she earn for a 44-hour week?

Answer ____________

2. Eric is a computer programmer that earns a yearly salary of $35,400. What is his weekly salary?

Answer ____________

3. Norma sells sports equipment. She earns a salary of $240 per week plus commission of 3% on all sales. One week she had sales of $3500. How much did she earn that week?

Answer ____________

4. In one pay period, Jack had $562.30 in regular earnings. His employer withheld $124.30 for taxes and $47.15 for insurance. What was Jack's take-home pay for that pay period?

Answer ____________

5. Henry's disability insurance pays 70% of his salary for up to 26 weeks. If his weekly salary is $450 and he is disabled for 20 weeks, how much disability pay will he get?

Answer ____________

6. Carrie has a monthly retirement income of $400 from a pension, $465 from social security, $375 from investments, and $200 from a part-time job. What is her total monthly retirement income?

Answer ____________

7. Complete the line graph for the table.

Tessoni's Windows Spring & Summer Sales
(Thousands of dollars)

April	45
May	70
June	75
July	55
Aug	50
Sept	30

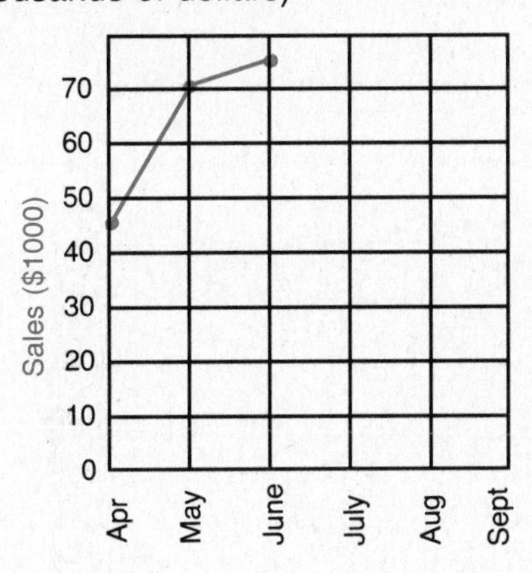

8. Ralph works part-time and has a yearly income of $5600, including a salary of $3750, and tips of $1850. What percent of his income was from his salary? What percent of his income was from his tips?

Answer ____________

Make a circle graph to show his income.

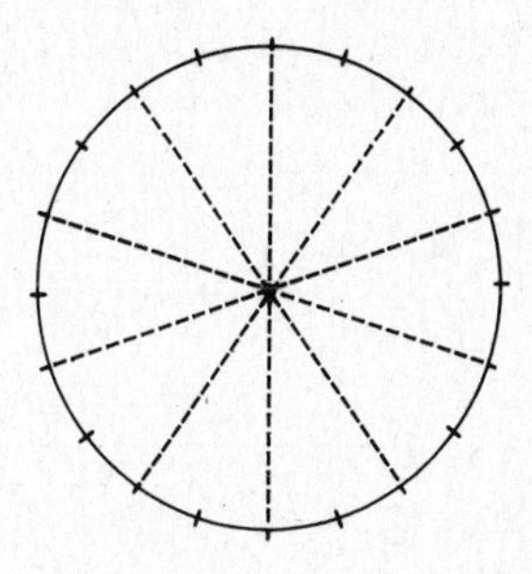

9. Look at the graph in Exercise 7. Between what 2 months did the sales increase the most?

Answer ____________

10. A store owner bought hats for $6.50 each and sold them for $12. The overhead was $2.50. What was the profit on each hat?

Answer ____________

UNIT 4 HANDLING MONEY

Comparing Prices

"A penny saved is a penny earned." A wise shopper can save money by comparing prices at different stores and by buying at discount prices. A **discount** is an amount subtracted from the regular price, often a percent of the regular price.

Some items are not priced individually. When comparing prices you may need to find the **unit price,** or the cost for one. If the cost does not divide evenly, round up to the next cent.

EXAMPLE Store A has scarves that are normally $4 each on sale at a 20% discount. Store B has the same type of scarves at 3 for $10. What is the price for a scarf at each store? Which store's price for a scarf is less?

STORE A
Discount: 20% of $4
$0.2 \times \$4 = \$0.80$
Sale price: $\$4.00 - 0.80 = \$3.20$

STORE B
Find the unit price by dividing $10 by 3.
$3)\overline{\$10.00}$ = $3.333 = $3.34

Since $3.20 is less than $3.34, Store A's price is less.

Find the amount of discount and the discount price of each item at Store A.

	Item	Reg. Price	Rate of Discount	Amount of Discount	Discount Price
1.	Camera	$32.50	10%	*$3.25*	
2.	Shoes	$39.50	15%		
3.	Bath towel	$10.00	25%		
4.	Shirt	$22.00	20%		
5.	File box	$8.80	50%		
6.	Radio	$12.49	$33\frac{1}{3}\%$		

7. The price of bath towels at Store B is 3 for $21. How much is the unit price? Is the price more or less than the sale price from Exercise 3?

Answer ______________________

8. The price of file boxes at Store B is 5 for $19.00. How much is the unit price? Is the price more or less than the sale price from Exercise 5?

Answer ______________________

9. The price of shirts at Store B is 2 for $40. How much is the unit price? Is the price more or less than the sale price from Exercise 4?

Answer ______________________

10. The regular price of shoes is $45. If the shoes are on sale at a discount of 30%, what is the sale price? Is the price more or less than the sale price from Exercise 2?

Answer ______________________

Using a Cash Account

A **cash account** is an itemized statement of income and expenses. To start a cash account, first label the columns on a ledger book or spreadsheet. Include columns for date, description, receipts, and different types of expenses such as food, clothing, housing, transportation, savings & giving. Label one column OTHER or MISC for miscellaneous expenses such as recreation. Write the current balance in the last column.

To use the cash account, list the date and description of each receipt or expense during the month and list the amount in the appropriate column. Add or subtract to find the balance. At the end of each month, find the total of each column. Then start a new page for the next month.

Study the cash account shown below. Then write each of these amounts in the cash account, finding the new balance each time.

1. 1/8 Shoes $43.78

2. 1/10 Gasoline $14.50

3. 1/12 Utility bills $85.79

4. 1/15 Car payment $257.00

5. 1/16 Groceries $78.11

6. 1/19 Wages $598.00

7. 1/22 Contributions $40.00

8. 1/24 Game tickets $16.40

9. 1/27 Savings $50.00

10. 1/30 Rent $540.00

			Expenses					Balance
Date	**Description**	**Receipts**	**Food & Clothes**	**Housing & Utilities**	**Transportation**	**Savings & Giving**	**Misc**	**$240.90**
1/2	Groceries		58.09					182.81
1/5	Wages	570.00						752.81
1/8	Contribution					37.00		715.81
1/8	*Shoes*		*43.78*					*672.03*
	TOTALS							

11. For the cash account above, find the total for the receipts and for each column of expenses.

12. Find the total of all the expenses. Is this amount more or less than the receipts?

Answer ______________

Preparing a Monthly Budget

A **budget** is a plan of how and in what amount income is to be spent. Whether you are choosing an apartment or buying new clothes, having a budget helps you know how much you can afford to spend.

To prepare a budget, you should look at your current spending records and estimate the amount needed for each budget category. Make sure to include savings and giving, and to allow for occasional expenses such as car insurance and repairs. Add the amounts budgeted for each category and compare this to your income (take-home pay). If the expenses are more or less than the income, **balance** the budget by adjusting the amounts for certain categories.

1. The list at the right shows Natalie's expenses for an average month. Find the total.

Answer ____________

NATALIE'S EXPENSES

Food & clothes	**$220**
Housing & utilities	**$430**
Transportation	**$90**
Savings & giving	**$100**
Miscellaneous	**$120**

2. Natalie's monthly income is $1070. Are the expenses less or more than the income?

Answer ____________

3. Natalie can balance her budget by putting more money in savings. How much more per month can she save?

Answer ____________

4. Natalie wants to buy a car. If she does, her transportation costs increase to $240 per month. Can she afford to buy a car without reducing other amounts in the budget?

Answer ____________

5. Lavon's income is $1300 per month. Here is his monthly budget except for food and clothes. To balance the budget, how much can he spend for food and clothes?

Housing & utilities	$280
Transportation	$320
Savings & giving	$130
Miscellaneous	$170

Food & clothes $ ____________

6. Edna Liebman is retired and has an income of $680 per month. Here is her monthly budget except for miscellaneous expenses. To balance the budget, how much can she spend for miscellaneous expenses?

Food & clothes	$160
Housing & utilities	$190
Transportation	$100
Savings & giving	$110

Miscellaneous $ ____________

7. Chun Lee buys clothing during January and July when the prices are lowest. If he spends a total of $420, how much should he set aside each month for clothing?

Answer ____________

8. The Alpert family takes a vacation each year that costs about $1600. What amount should be saved or set aside each month for vacation expenses?

Answer ____________

Family Budgets

Circle graphs are helpful for showing the portion, or percent, that is budgeted for each category. The two circle graphs show monthly budgets for the Moore family and the Pasko family.

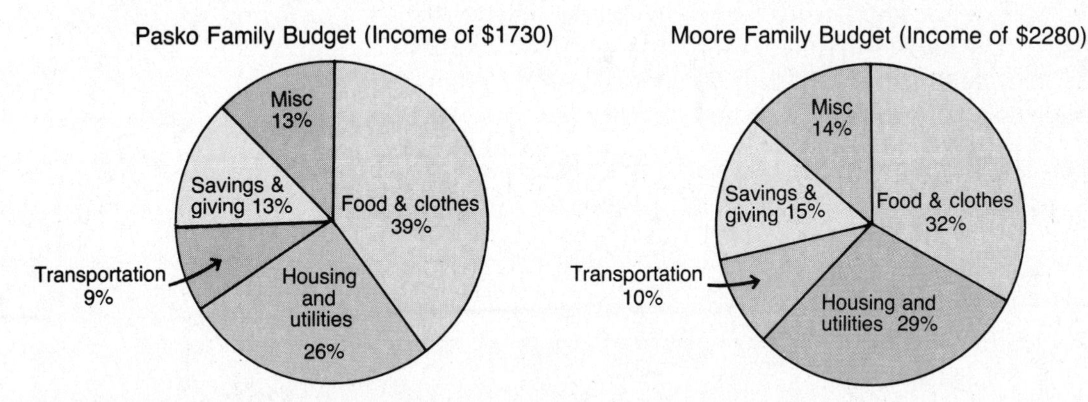

Which family has a higher percent budgeted for food and clothes each month?

Since 39% is greater than 32%, the Pasko family's budget includes a greater percent for food and clothes. However, the *amount* budgeted on food depends on the income of each family.

Find the amount that each family has budgeted for each category. (Hint: Write the percent as a decimal and multiply by the income.)

1. Pasko Family (Income of $1730 per month)

Food & clothes	$1730 \times 0.39 =$ ________
Housing & utilities	________
Transportation	________
Savings & giving	________
Miscellaneous	________
TOTAL	________

2. Moore Family (Income of $2280 per month)

Food & clothes	$2280 \times 0.32 =$ ________
Housing & utilities	________
Transportation	________
Savings & giving	________
Miscellaneous	________
TOTAL	________

3. Which family budgets a greater amount each month for food and clothes? Use your answers from Exercises 1 and 2.

Answer ________

4. Which family budgets a greater amount each month for transportation? Use your answers from Exercises 1 and 2.

Answer ________

5. Suppose each family saves 5% of its income each month. How much does each family save?

Answer ________

6. Suppose each family spends $100 a month for recreation. What percent does each family spend for recreation?

Answer ________

Checking Accounts

Many people write checks to pay for purchases or bills. Using a checking account is safer than paying cash, and both you and the bank will have records of how much is spent.

Opening a checking account at a bank requires a deposit and your signature on a signature card. The bank will give you some checks and deposit slips, then order printed checks that you will receive in about a week. Banks usually charge a monthly fee for checking accounts.

The amount of a check is written in both words and numbers. If there is a difference in these amounts, the bank will honor the amount written in words.

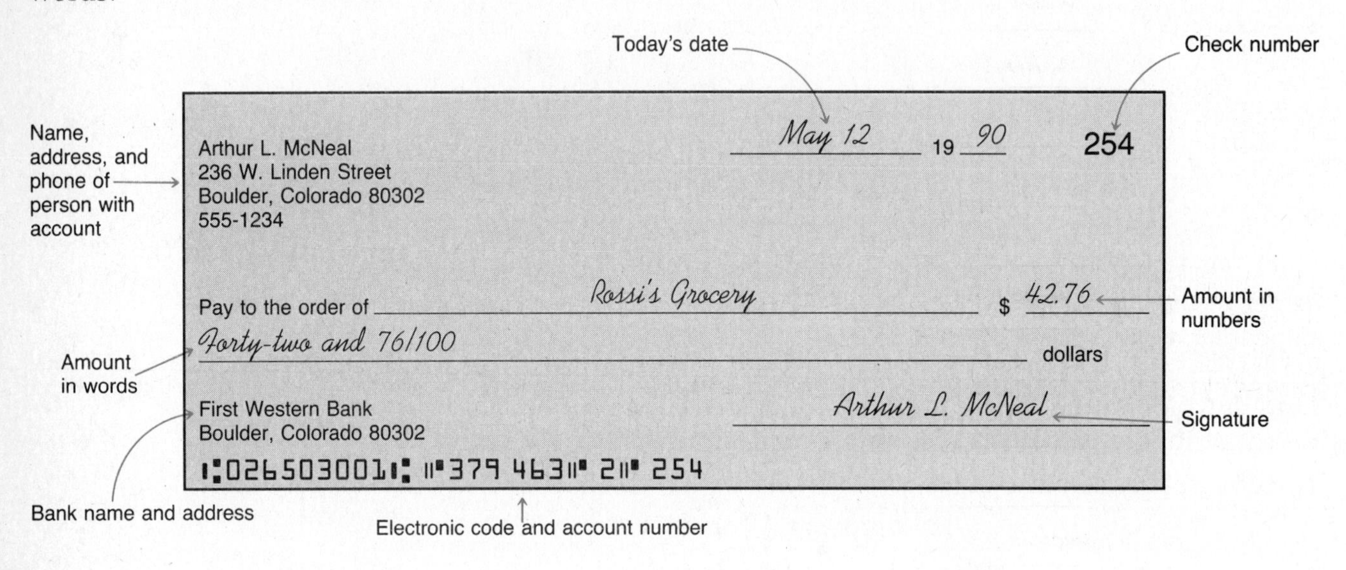

Write each amount in words as it would appear on a check.

1. $15.20 ______________________ **2.** $40.10 ______________________

3. $243.08 ______________________

4. $106.90 ______________________

5. Suppose you must pay $75.10 to Long's Hardware Store. Complete the check below for that amount using today's date and your name.

____________ 19___ 158

Pay to the order of ______________________ $ ________

______________________ dollars

First Western Bank
Boulder, Colorado 80302 ______________________

Check Registers

A **check register** is sent to you with your checks to keep track of the checks you write, deposits you make, and the balance you have left in your checking account. This register page shows two check payments and one deposit. After each entry, you add or subtract to find the current balance in the account.

Number	Date	Transaction	Payment/Debit		Deposit/Credit		Balance		
							724	59	← Read this as $724.59.
457	5/7	Weinstein's Nursery	38	82			− 38	82	
		(garden supplies)					685	77	← New Balance
458	5/10	Mrs. Phyllis Kortney	375	00			−375	00	
		(rent)					310	77	← New Balance
	5/15	Deposit			872	36	+872	36	
		(paycheck)					1183	13	← New Balance

1. Find the three new balances in the right column.

Number	Date	Transaction	Payment/Debit		Deposit/Credit		Balance	
							974	82
145	12/2	South City Edison	83	70			83	70
		(electric bill)						
146	12/8	World Cable	14	57			14	57
		(Cable TV)						
	12/9	Deposit			475	80	475	80
		(Paycheck)						

2. On July 13, Maria deposited her paycheck for $650.25. Enter this amount in the Deposit/Credit column. Then complete all entries in the Balance column.

Number	Date	Transaction	Payment/Debit		Deposit/Credit		Balance	
							1075	65
175	7/10	Alfred's Furniture	153	58			153	58
		(kitchen table)						
176	7/12	New Jersey Bell	34	79			34	79
		(telephone bill)						
	7/13	deposit						
		(paycheck)						

PROBLEM-SOLVING STRATEGY

Make a List

Sometimes it helps to make a list to solve a problem. When you put information in a list in an organized way, it is easy to see how to solve the problem.

Read the problem.

When you open a checking account, you get to choose a design for the checks and a color for the checkbook cover. Suppose you can choose a design with pictures of the seashore, animals, or hearts. The color for the cover can be red, blue, green, or tan. How many different combinations of check design and color are possible?

Make a list.

Label the designs S, A, and H. Label the colors R, B, G, and T. Then make a list of possibilities. Be sure the list is organized carefully.

Seashore	*Animals*	*Hearts*
S-R	A-R	H-R
S-B	A-B	H-B
S-G	A-G	H-G
S-T	A-T	H-T

Solve the problem.

Count the number of combinations listed. There are 12 different combinations of design and color.

Make a list. Use your list to solve the problem.

1. The checks in Dwayne's checkbook are numbered from 1051 to 1075. How many of the check numbers are odd numbers?

Answer ______________________

2. The Underwood family is planning a vacation. They are going to a museum, to an amusement park, and to a zoo. They are deciding the order in which to visit each place. How many different orders are possible?

Answer ______________________

Make a list. Use your list to solve the problem.

3. The checks in Ralph's checkbook are numbered from 530 to 550. How many of the checks in the book have two digits that are the same? (Hint: 550 has two fives.)

Answer ____________

4. The checks in Rachel's checkbook are numbered from 777 to 800. How many of the check numbers have digits with a sum greater than 20? (Hint: The digits of 777 have a sum of 7 + 7 + 7 = 21.)

Answer ____________

5. A snack bar sells small glasses of juice for $0.75 and large glasses of juice for $0.95. Cold sandwiches cost $1.10 and hot sandwiches cost $1.30. If a person buys one sandwich and one glass of juice, how many different total costs are possible?

Answer ____________

6. The Spencer family is going shopping. They need to stop at the bookstore, grocery, shoestore, and pharmacy. They are deciding the order in which to visit each place. The grocery must be the third or fourth stop. How many different orders are possible?

Answer ____________

7. The balance in Andrea's checkbook was $450.55. When she copied the balance, she accidentally reversed two adjacent digits. How many different amounts could Andrea have written?

Answer ____________

8. The balance in Tony's checkbook was an exact number of dollars between $110 and $180. The last digit of the number was 4. How many such amounts are possible?

Answer ____________

Savings Accounts: Simple Interest

When you deposit money in a bank, you are allowing the bank to use the money to earn more money for the bank. The bank pays you for this. The amount that they pay you is called **interest.** The amount you deposit is called the **principal.** The **rate** is the percent of the principal that you earn in one year. The formula used to find the interest is **$I = p \times r \times t$.**

> **$I = p \times r \times t$**
> **I** is the amount of interest.
> **p** is the principal, or amount of money you deposited.
> **r** is the rate, or percent of interest per year.
> **t** is the time in terms of years.

EXAMPLE Suppose you deposit $600, and the rate is 3%. How much interest would you earn in 3 months?

Since the rate is based on years, 3 months is $\frac{3}{12}$ or $\frac{1}{4}$ of a year.

You can use a fraction or a decimal for the number of years.

$I = p \times r \times t$ $I = p \times r \times t$

$I = \$600 \times 0.03 \times \frac{1}{4}$ **or** $I = \$600 \times 0.03 \times 0.25$

$I = \$4.50$ $I = \$4.50$

Complete each row of the chart.

	Principal	Rate	Time	Time in Years	I = p X r X t
1.	$500	7%	6 months	*$\frac{1}{2}$ or 0.5 year*	*$500 X 0.07 X 0.5 =*
2.	$1000	5%	1 year		
3.	$900	2%	4 months		
4.	$600	3.5%	9 months		
5.	$1600	4%	3 years		
6.	$6000	6.75%	5 years		

7. Willis Brooks deposited $400 in a savings account at 3% interest. How much interest will he earn in 9 months?

Answer ______________

8. Elena Ramaro withdrew her savings at the end of $\frac{1}{2}$ year. If she had deposited $800 at 2.5%, how much interest did she get? What was the amount she withdrew?

Answer ______________

9. Mark Adams has $300 in a savings account at 2% interest. How much interest will he earn in 3 months?

Answer ______________

10. If you deposit $1000 at 5% interest, how much interest will you earn each month? How long will it take to earn $100 interest?

Answer ______________

Savings Accounts: Compound Interest

Patricia deposited $100 in a savings account at 5% interest. At the end of 1 year, she will have $5 in interest. If Patricia leaves the $5 in the account, her new principal will be $105. During the second year, the bank will pay interest on the original principal *plus* the interest that has been added to the account. Interest paid in this way is said to be compounded annually (once a year).

EXAMPLE How much interest will be in Patricia's account at the end of the second year?

Interest for first year:

p	×	r	×	t	=	I
$100	×	0.05	×	1	=	$5.00

Account balance at end of first year:

Prior balance		Interest		New balance
$100	+	$5	=	$105

Interest for second year:

$105 × 0.05 × 1 = $5.25

Account balance at end of second year:

$105 + $5.25 = $110.25

At the end of the second year the balance is $110.25.

Solve.

1. George has $250 in the savings bank and gets 5% interest compounded annually. How much interest will he get for the second year?

Answer ______________________

2. Madge has $300 in her savings account and is paid 6% interest compounded annually. How much will she have in her account at the end of the second year?

Answer ______________________

3. Pat has $500 in a savings account which earns 10% interest. How much will he have in the bank at the end of the second year if the interest is compounded annually?

Answer ______________________

4. Evita has $170 in a savings account. She is paid 6% interest compounded annually. How much will she have in her account at the end of the second year?

Answer ______________________

5. Ruth Russell deposited $350 in a savings account. She received 6% interest compounded semiannually (twice per year). How much interest did the account earn during the second half of the year? (Hint: Use 0.5 for the time.)

Answer ______________________

6. Frank has $500 in a savings account and is paid 5% interest compounded semiannually. How much will he have in the account at the beginning of the second year?

Answer ______________________

Savings Accounts: Daily Compounding

Because computers can make complex computations quickly, many banks offer interest compounded daily rather than yearly. With yearly compounding, a deposit of $100 at 2% earns $2 in 365 days. However, if interest is compounded daily, the deposit will earn more than $2 in a year. This is because interest is added to the account each day, and this interest then earns interest.

Stated Rate	Effective Annual Rate (Daily Compounding)
2%	2.02%
3%	3.05%
4%	4.08%
5%	5.13%
6%	6.18%
7%	7.25%
8%	8.32%

If the **stated rate** is 2% and interest is compounded daily, the **effective rate** for one year is 2.02%. With daily compounding, a deposit of $100 would earn $2.02 interest in one year. The table shows effective interest rates for other stated rates.

EXAMPLE The stated rate of interest on a savings account is 3%. If interest is compounded daily, how much annual interest would be earned on a deposit of $600?

The effective annual rate is 3.05%.

Principal		Effective Annual Rate		Annual Interest
$600	×	0.0305	=	$18.30

Complete each row of the chart to find the annual interest earned. The interest is compounded daily.

	Principal	Stated Rate	Effective Annual Rate	Annual Interest
1.	$1000	5%	*5.13% or 0.0513*	*$1000 × 0.0513 =*
2.	$5000	2%		
3.	$850	7%		
4.	$1130	3%		
5.	$280	8%		
6.	$585	4%		

7. Suppose you deposit $1000 and earn simple interest at 5% for one year. How much interest would you earn?

Answer ____________________

8. Compare your answers to Exercises 1 and 7. How much more interest would you earn when interest is compounded daily?

Answer ____________________

Calculator Applications

The diagram below leads to a shortcut that is helpful when calculating compound interest.

Suppose you need to find the new balance after 5% interest is added. The prior balance is 100% of itself. The interest is 5% of the prior balance. So, the new balance is 100% + 5%, or 105% of the prior balance.

Prior balance 100%	Interest 5%
New balance 105%	

EXAMPLE Suppose interest is compounded quarterly and the rate is 6% per year or 1.5% per quarter. An account has $470 at the beginning of a year. How much is in the account at the end of the year?

Note that 100% + 1.5% is 101.5%. So, to find the balance at the end of each quarter, multiply the prior balance by 101.5% or 1.015.

Quarter	Key Sequence		New Balance	
1st	470 [×] 1.015 [=]	477.05	$477.05	(Note: Do not clear the display.)
2nd	[×] 1.015 [=]	484.20575	$484.21	
3rd	[×] 1.015 [=]	491.46884	$491.47	
4th	[×] 1.015 [=]	498.84087	$498.84	

The balance at the end of the year is $498.84.

Solve. Use a calculator.

1. Suppose the prior balance was $200. By what number would you multiply to find the new balance after 2.5% interest is added? What is the new balance?

Answer ________________

2. Suppose the prior balance was $350. By what number would you multiply to find the new balance after 5.2% interest is added? What is the new balance?

Answer ________________

3. A savings account earns 5% interest compounded quarterly. If the account has $600 at the beginning of the year, how much is in the account at the end of the year? (Hint: Since the quarterly interest rate is 1.25%, multiply by 1.0125 four times.)

Answer ________________

4. Suppose $100 is deposited in a savings account that has an effective annual rate of 5.3%. How much money would be in the account in 5 years? (Hint: Multiply by 1.053 five times.)

Answer ________________

U.S. Savings Bonds

Series EE Savings Bonds can be purchased with face values of $50, $75, $100, $200, $500, $1000, $2000, and $10,000. The cost of these bonds is one-half the face value. For example, the cost of a $500 bond is $250. Each bond **matures,** or reaches its face value, in 8 to 12 years.

The bond can be **redeemed,** or cashed in, any time after 6 months. The **redemption value** of the bond includes interest for the time the bond was owned. The actual interest rate may change every 6 months, but a minimum rate of about 6% is guaranteed. The guaranteed redemption values for a $50 Series EE Savings Bond are shown in the table.

Series EE $50 Savings Bonds	
Time	**Redemption Value**
1 yr	$26.08
2 yr	$27.40
3 yr	$29.00
4 yr	$31.06
5 yr	$33.60
6 yr	$35.66
7 yr	$37.82
8 yr	$40.12
9 yr	$42.58
10 yr	$45.16

EXAMPLE What is the redemption value of a $1000 bond after 5 years?
The redemption value of any bond is based on the redemption value of a $50 bond.

Value of $50 bond after 5 years		Number of $50 bonds equal to $1000 bond (1000 ÷ 50)		Redemption value of $1000 bond after 5 years
$33.60	×	20	=	$672.00

Solve. Use the table of redemption values.

1. Alex buys four $200 savings bonds. How much does he pay for them?

Answer ____________

2. Wanda buys five $75 savings bonds. How much does she pay for them?

Answer ____________

3. Norm Pulaski buys a $100 savings bond. What will be the redemption value in 4 years?

Answer ____________

4. Agnes Barclay buys ten $50 savings bonds. What will be their total worth in 5 years?

Answer ____________

5. Marlene spent $2500 for bonds. What is the minimum that the bonds will be worth in 10 years?

Answer ____________

6. If Marlene puts $2500 in savings at 5% interest compounded daily, she would have $4072 in 10 years. Which is a better investment, a savings account at 5% or Series EE bonds?

Answer ____________

Certificates of Deposit

FIRST COUNTY BANK CERTIFICATES OF DEPOSIT MINIMUM DEPOSIT $1,000	
TERM	RATE
3 months	2.68%
6 months	2.89%
1 year	3.04%
3 years	3.28%
5 years	5.23%

Most banks offer certificates of deposit as well as savings accounts. A **certificate of deposit,** or **CD,** earns a higher interest rate than a savings account. This is because you must deposit a minimum amount and promise to leave the money in the bank for a certain period of time. If the money is withdrawn before the certain period of time is over, you will be charged a penalty.

On CDs for periods longer than 32 days, each individual bank can set the minimum deposit required and the interest it will pay. Usually the bank will pay higher rates for money left on deposit for longer times. The table above shows the rates one bank was offering during a certain week.

EXAMPLE A customer invests $2400 in a 6-month CD that has a yearly interest rate of 2.89%. How much interest will the customer receive at the end of 6 months?

Principal	×	Rate	×	Time	=	Interest
$2400	×	0.0289	×	0.50	=	$34.68

Complete each row of the chart.

	Principal	Rate	Time	Time in Years	I = p × r × t
1.	$5000	2.68%	6 months	$\frac{1}{2}$ or 0.5	$5000 × 0.0268 × 0.5 =
2.	$1500	3.28%	3 years	3	
3.	$3500	2.89%	3 months	$\frac{1}{4}$ or 0.25	
4.	$7200	3.04%	3 years		
5.	$2000	5.23%	5 years		
6.	$2500	4.75%	1 year		

7. José Ramirez invests $5000 in a 6-month CD that earns 3.10%. How much interest will he receive at the end of 6 months?

Answer ____________________

8. Willa Johnson buys a 9-month CD with an interest rate of 2.65%. If she invests $2000, how much interest will she get at the end of 9 months?

Answer ____________________

9. Margaret Mabry has two $3000 CDs, each for 6 months at 2.62%. How much interest will she get from her CDs? How much money will she have in all at the end of the 6 months?

Answer ____________________

10. Alec Morgan invests $4000 in a 1-year CD at 3.04%. He withdraws the money at the end of 6 months and is charged a penalty of 3 months interest. How much interest does he receive?

Answer ____________________

Stocks

Many people invest money by purchasing **stocks.** A person who owns a share of stock is a partial owner of the company or business that issues the stock. The company uses money from the sale of stocks to help run the company. If the company makes a profit during the year, owners of stock are given **dividends,** or portions of the profit.

Suppose a business issued 5000 shares of stock. Jan Karlin bought 1000 shares. If the business gives out $15,000 in dividends, how much will Jan receive in dividends on her shares?

Jan Karlin owns $\frac{1000}{5000}$ or 0.20 or 20% of the stock. She will get 20% of $15,000, or $3000 in dividends.

Solve.

1. Myron Nelson bought stock in a manufacturing company. The first year he was paid a dividend of $5.25 per share on the 100 shares he had bought. How much did he receive in all?

Answer ____________________

2. Myron Nelson bought 90 shares of stock at $100 each. Three years later, he sold the stock for $95 per share. How much money did he lose?

Answer ____________________

3. Rose Wentworth owns 50 shares of stock in an oil company. The total number of shares of stock owned by all the stockholders in the company is 5000. What fraction of the shares does Rose own?

Answer ____________________

4. If the oil company declares dividends totaling $25,400, how much will Rose Wentworth receive in dividends?

Answer ____________________

5. Bob's mother gave him 10 shares of stock in a local company. She paid $52 per share for them. At the end of the first year, the company paid a dividend of $2.90 per share. How much did Bob receive in dividends?

Answer ____________________

6. The second year, Bob received $3.25 per share in dividends. How much more money did Bob receive the second year than the first year?

Answer ____________________

Single-Payment Loans

At a bank, some of the money that is deposited by customers in their savings accounts is used for loans to other customers. The bank *earns* interest from borrowers, but *pays* interest to depositors. The bank, of course, must charge a higher rate of interest on loans than it pays on savings accounts. Otherwise it could not make enough money to stay in business.

A **single-payment** loan is one that is repaid with one payment after a specified period of time. Interest on a single-payment loan is figured using the simple interest formula, $I = p \times r \times t$. If the loan is for more than a year, the customer may be required to pay the interest each year.

EXAMPLE Jason Williams borrowed $560 from a bank to remodel his family room. The loan was for 4 months at 12% interest. How much interest did he pay? Remember to write the time as a number of years.

Principal		Rate (12%)		Time $\left(\frac{4}{12}\right)$		Interest
$560	×	0.12	×	0.333	=	$22.38

Complete each row of the chart.

	Principal	Rate	Time in Months	Time in Years	p × r × t = Interest
1.	$500	8%	6	$\frac{1}{2}$ or 0.5	$500 × 0.08 × 0.5 =
2.	$820	12%	12		
3.	$1300	9%	3		
4.	$1530	10%	4		
5.	$2100	11%	18		
6.	$5900	10%	9		

7. Kathy borrowed $250 at 18% interest for 6 months. How much interest did she pay?

Answer ______________

8. Molly Evans borrowed $1250 at 13% interest for 6 months. How much interest would she owe on the loan?

Answer ______________

9. Harold Bowland borrowed $500 from the bank at 11% interest for 8 months. Including interest, how much did he have to repay?

Answer ______________

10. A rancher borrowed $6000 from the bank for 4 months and agreed to pay 12% interest. Including interest, how much did he have to repay?

Answer ______________

PROBLEM-SOLVING STRATEGY

Use Estimation

Many problems can be solved by estimation. Often, you do not need an exact answer to solve a problem. An estimate is found by rounding some or all of the numbers and then doing mental math.

Read the problem.

Ernest deposited $380 in a savings account that has an interest rate of 5.2%. About how much interest will be earned in 2 years?

Identify the important facts.

P = $380, r = 5.2% or 0.052, t = 2

Round.

Round $380 to $400 and round 5.2% to 5%.

Solve the problem.

Principal		Rate		Time		Interest
$400	×	0.05	×	2	=	$40

In two years, about $40 interest will be earned.

Use estimation to solve each problem.

1. Norman deposited $865 in a savings account that has an interest rate of 5.8%. About how much interest will be earned in 6 months?

Answer ____________________

2. Winona had a balance of $614 in her checking account. Then she deposited $385 and wrote a check for $423. Estimate the new balance.

Answer ____________________

3. Here are the amounts someone spent for groceries during a 3-week period.

$93 $78 $95

About how much money should be budgeted each week for groceries?

Answer ____________________

4. Hubert deposited $387 in a savings account at 7% interest. At this interest rate, the deposit will double in about 10 years. About how much money will be in the account in 10 years?

Answer ____________________

Use estimation to solve each problem.

5. Sarah deposits $723 in a savings account at 2.5% interest. At this interest rate, the deposit will double in about 28 years. About how much money will be in Sarah's account after 28 years?

Answer ______________________

6. To estimate the number of years it will take for an investment to double in value, divide 70% by the interest rate. About how many years will it take for an investment to double if the interest rate is 4.9%?

Answer ______________________

7. Joseph invests $1600 in a 2-year certificate of deposit that pays 3.12% interest. About how much interest will Joe receive on his investment?

Answer ______________________

8. Della invests $3200 in a 6-month certificate of deposit that pays 2.89% interest. About how much interest will she receive on her investment?

Answer ______________________

9. The Rossi family has a monthly budget of $2300. Last month 19.8% of the budget was spent for food. About how much money was spent for food?

Answer ______________________

10. The Tyler family has a monthly budget of $1780. Last month 32% of the budget was spent for housing and utilities. About how much money was spent for housing and utilities?

Answer ______________________

11. Elinor owns 39 shares of stock. She received dividends of $2.05 per share. About how much did she receive in all?

Answer ______________________

12. Lewis owns 806 shares of stock. He received dividends of $0.87 per share. About how much money did he receive in all?

Answer ______________________

Installment Loans: Monthly Payments

Recall that a single-payment loan is repaid as a lump sum with interest. Most people find it easier to repay a loan with monthly payments, called **installments.** The table at the right can be used to find the amount of each monthly payment.

Businesses sometimes allow customers to pay for purchases with monthly installments. A lender or business that offers installment loans must inform borrowers of the interest rate, called the **annual percentage rate** or APR.

Monthly Payment per $100			
APR	6 mo	12 mo	18 mo
8%	$17.06	$8.70	$5.92
10%	$17.16	$8.79	$6.01
12%	$17.26	$8.89	$6.10
14%	$17.36	$8.98	$6.19
16%	$17.46	$9.07	$6.29
18%	$17.56	$9.16	$6.39

EXAMPLE Martha Conners bought a stereo for $940. She made a $140 down payment. The store offered to lend her the balance. She decides to repay the loan in 18 monthly installments. If the APR for the loan is 18%, how much is each monthly payment?

Cost		Down payment		Amount of loan
$940	−	$140	=	$800

Payment per $100		Loan in hundreds		Monthly payment
$6.39	×	8	=	$51.12

Each monthly payment is $51.12.

Find the monthly payment. Use information from the table of monthly payments.

	Amount	APR	Months	Payment
1.	$100	10%	6	
2.	$400	14%	6	
3.	$900	18%	12	
4.	$1400	12%	18	

	Amount	APR	Months	Payment
5.	$350	8%	18	
6.	$620	16%	12	
7.	$1700	14%	12	
8.	$4900	10%	18	

9. Mrs. Newhouse bought a tent for $150. She made a down payment of $50 and paid the balance in 12 monthly installments. If the APR was 16%, how much was each installment?

Answer ______________________

10. Matt bought a $2400 used car with a down payment of $800. He paid the balance in 6 monthly installments. If the APR was 12%, how much was each installment?

Answer ______________________

Installment Loans: Total Interest

To find the total interest on an installment loan, multiply the number of months by the amount paid each month. Then subtract the amount borrowed.

EXAMPLE How much interest is paid on a $3000 installment loan at 18%? The loan is to be repaid in 6 installments. The table on page 92 shows that the payment per $100 is $17.56.

Monthly payment: 30 (Loan in hundreds) × $17.56 = $526.80
Total of 6 payments: 6 × $526.80 = $3160.80
Total interest: $3160.80 − $3000 = $160.80

Complete each row of the chart.

	Amount of Loan	Interest (APR)	Number of Months	Monthly Payment	Total of Payments	Total Interest
1.	$200	14%	6	$34.72		
2.	$500	12%	12	$44.45		
3.	$1100	16%	18	$69.19		
4.	$2300	10%	12	$202.17		

Part of each installment is interest and the rest is applied to the principal, or amount owed. Because the amount owed on the loan decreases each month, the interest each month also decreases. This is shown by the payment schedule at the right. The amounts shown have been rounded.

Payment Schedule
6-Month Installment Loan of $3000, 18% Interest

Payment Number	Payment	Interest Paid	Principal Paid	Balance
				$3000
1	$527	$45	$482	$2518
2	$527	$38	$489	$2029
3	$527	$31	$496	$1533
4	$527	$24	$503	$1030
5	$527	$16	$511	$519
6	$527	$8	$519	0

Answer each question about the payment schedule at the right.

5. For each payment, what is the sum of the interest paid and the principal paid?

Answer ____________

6. What is the sum of the amounts in the column for *Principal Paid?*

Answer ____________

7. What is the sum of the amounts in the column for *Interest Paid?*

Answer ____________

8. How much interest would be paid on a single-payment loan of $3000 at 18% for 6 months? Will less interest be paid for a single-payment loan or an installment loan?

Answer ____________

Credit Cards

A credit card allows you to pay less than the full amount you owe each month. But when you do not pay the full amount, you must pay a finance charge. This charge is found by multiplying the unpaid balance by the monthly interest rate, which is $\frac{1}{12}$ of the annual interest rate.

EXAMPLE Arnold's unpaid balance on his credit card is $340. This month he made a payment of $70 and charged purchases totaling $175.78. The finance rate is 18% per year or 1.5% per month. What is his new balance?

Old balance		Payment		Balance after payment
$340	−	$70	=	$270

Rate		Balance after payment		Finance charge
0.015	×	$270	=	$4.05

To find the new balance, add the finance charge, $4.05, and new purchases, $175.78, to the balance after payment.

New balance: $270 + $4.05 + $175.78 = $449.83

Solve.

1. Mark's unpaid balance for last month was $258. He made a payment of $50 and did not use his credit card this month. What is the amount on which he will be charged a finance charge?

Answer ____________

2. If the finance rate on Mark's credit card is 2% per month, what will be his finance charge on the unpaid balance?

Answer ____________

3. If a department store credit card has a finance rate of 1.5% per month, what will be the finance charge on an unpaid balance of $560.80?

Answer ____________

4. Angela has an unpaid balance of $320 on her credit card. The finance rate is 1.5% per month. If she makes a payment of $50 and does not make any new purchases, what will be her new balance?

Answer ____________

5. Mr. Martinelli's unpaid balance is $300. If he makes a payment of $75 and buys an airplane ticket for $220, what will be his new balance? The finance rate is 1% per month.

Answer ____________

6. Dr. Wong's credit card has a finance rate of 2% per month. Her unpaid balance is $1600. If she makes a payment of $40 and buys a table for $150, what will be her new balance?

Answer ____________

Unit 4 Review

Solve.

1. Phyllis has a yearly income of \$26,000. She budgets 15% for clothing. How much will she have for clothing?

Answer ____________

2. One month, Georgia earned \$350. She spent \$125 for rent, \$45 for food, and \$70 for miscellaneous items. How much did she have left to put into her savings account?

Answer ____________

3. Agnes has a balance in her checking account of \$450.70. If she writes a check for \$35.15 and makes a deposit of \$120.40, what will her new balance be?

Answer ____________

4. Bill deposited \$520 in a savings account that pays 6% simple interest per year. How much interest will he earn in 1 year?

Answer ____________

5. Frances puts \$1200 into a savings account that pays 6% interest, compounded annually. How much will she have in the account in 2 years?

Answer ____________

6. Barbara buys three \$100 U.S. savings bonds for \$50 each. A \$100 bond will be worth \$90.32 in 10 years. How much will Barbara's bonds be worth in 10 years?

Answer ____________

7. Bob has \$6000 in a 1-year CD at 7% interest. How much interest will he receive at the end of 1 year?

Answer ____________

8. A farmer borrowed \$13,000 for 6 months at 11% interest. How much interest will he pay on the loan?

Answer ____________

9. Ronnie borrowed \$1500 to buy a computer. He will repay the loan in 12 installments of \$134.70 each. How much interest will he pay on the loan?

Answer ____________

10. Susan has an unpaid balance on her credit card of \$450. If the finance rate is 2% per month, what is the finance charge on her balance?

Answer ____________

Buying a New Car: Sticker Price

Buying a new car requires careful consideration of various costs. At the factory, a sticker is put on a new car to show the suggested sales price. The **sticker price** includes the base price of the car plus the cost of special equipment (options) plus the cost of shipping the car from the factory to the dealer (destination charge). The costs for sales tax and license fees are *not* included on the sticker and are added later.

SPARKY (Compact Car)	
BASE PRICE	$8980
OPTIONS:	
Power windows	$265
Automatic transmission	$599
Air conditioning	$915
Power brakes	$84
Tinted glass	$75
Cruise control	$224
DESTINATION CHARGE	$378

A customer ordered a Sparky with power windows, air conditioning, and power brakes. Use prices from the chart to find the sticker price.

Option costs: $265 + $915 + $84 = $1264
Sticker price: $8980 + $1264 + $378 = $10,622

Solve. For Exercises 1 and 2, refer to the chart above.

1. A customer orders a Sparky with automatic transmission and air conditioning. What is the sticker price? (Hint: Don't forget the destination charge.)

Answer ______________

2. A customer orders a Sparky with all six of the options that are listed. What is the sticker price?

Answer ______________

3. A customer orders a minivan with these options: automatic transmission, $724; power steering, $329; and cruise control, $247. What is the total cost of options?

Answer ______________

4. The minivan from Exercise 3 has a base price of $11,941 and destination charge of $445. What is the sticker price?

Answer ______________

5. The base price of an Encore TZ is $9052. The customer has selected options that have a total cost of $713. The destination charge is $385. What is the sticker price?

Answer ______________

6. The base price of a conversion van is $13,108. The customer selects options that cost $859. The destination charge is $480. What is the sticker price?

Answer ______________

Buying a New Car: Dealer's Cost

The dealer's cost of buying the car from the factory is usually 75% to 80% of the sticker price. A customer might save money by offering a price that is between the sticker price and the dealer's cost.

Consumer magazines often give reports of the dealer's cost for a car. For the Sparky the dealer pays 78% of the base price, 75% of the cost of options, and all of the destination charge.

How much does the dealer pay for a Sparky with power windows, air conditioning, and power brakes? Would $9100 be a reasonable offer? For each step, round the amount to the nearest dollar.

Dealer's cost:

78% of base price = 0.78 × $8980 =	$7004
75% of options = 0.75 × $1264 =	948
Destination charge =	+ 378
	$8330

The offer of $9100 seems reasonable. However, if there is high demand for the car, the dealer may not agree to sell for less than the sticker price.

Solve. Round to the nearest dollar.

1. A dealer buys a small station wagon with no options. The dealer's cost is 78% of the base price of $9588, plus a destination charge of $285. Find the dealer's cost.

Answer ____________________

2. For the station wagon in Exercise 1, what is the difference between the sticker price and the dealer's cost?

Answer ____________________

3. The dealer's cost for a fullsize car includes 82% of the base price of $14,560 and 78% of the options cost of $2148. The dealer's cost also includes a destination charge of $357. What is the dealer's cost?

Answer ____________________

4. The dealer's cost for a midsize car includes 78% of the base price of $11,099 and 75% of the options cost of $1343. The dealer's cost also includes a destination charge of $277. What is the dealer's cost?

Answer ____________________

5. For the fullsize car in Exercise 3, the dealer accepts an offer that is 10% less than the sticker price. How much does the customer pay for the car?

Answer ____________________

6. For the midsize car in Exercise 4, the dealer accepts an offer that is $1300 less than the sticker price. How much does the customer pay for the car?

Answer ____________________

DRIVING A CAR

Buying a Used Car

A person who is buying a used car may save money by offering less than the asking price. This is because car dealers often set prices that are higher than what they expect to get. After a buyer makes an offer, the dealer probably will adjust the price to an amount between the buyer's offer and the dealer's original price. So, wise buyers often make offers that are less than the dealer's original price.

There are monthly publications, or guides, that can help you decide how much you might offer for a used car. These guides list cars by manufacturer, model, and year. Data such as the wholesale value and a suggested retail price may also be provided. Keep in mind that the condition of a particular car and the number of miles it has been driven can affect these prices. The popularity of a given model can also affect the prices. Most community libraries have these guides available in their reference sections. You should feel free to ask the librarian to help you find them.

Used Car Model	Wholesale Price	Suggested Retail Price
Accord	\$10,200	\$12,950
Bronco	\$15,100	\$18,850
Camaro	\$10,200	\$13,050
Ciera	\$8,875	\$11,500
Mustang	\$9,050	\$11,700
Shadow	\$8,450	\$9,575

A dealer is asking \$13,200 for the Camaro shown in the table. Rene offers \$1500 more than the wholesale price. The dealer lowers the asking price of the car by \$1,250. Because the car is in good shape, Rene accepts the offer. What was Rene's original offer and how much did she agree to pay for the Camaro?

Rene's offer:	\$10,200	+	\$1500	=	\$11,700
Agreement price:	\$13,200	−	\$1250	=	\$11,950

Solve. Use the prices shown in the chart.

1. Find the difference between the suggested retail price and the wholesale price for each model listed in the chart.

   Accord ___________ Bronco ___________

   Camaro ___________ Ciera ___________

   Mustang ___________ Shadow ___________

2. Use your answers for Problem 1. For which model is the difference between the wholesale price and the retail price the greatest?

   Answer ___________________

3. Suppose you believe that a 20% profit for the dealer is a fair amount. What price would you be prepared to pay for each model listed in the chart? [Hint: Multiply each wholesale price by 120%, or 1.2.]

   Accord ___________ Bronco ___________

   Camaro ___________ Ciera ___________

   Mustang ___________ Shadow ___________

4. Suppose you believe that a 10% profit for the dealer is a fair amount to give as your first offer. What would be your first offer for each model listed in the chart?

   Accord ___________ Bronco ___________

   Camaro ___________ Ciera ___________

   Mustang ___________ Shadow ___________

Renting a Car

If a person only needs a car occasionally, it may be less expensive to rent than to buy. A car rental company usually has daily or weekly rates. Some companies also charge an additional amount for the mileage. The chart at the right shows the rates at Carla's Car Rental.

Carla's Car Rental		
Type	**Daily Rate***	**Weekly Rate***
Compact	$26.95	$169.00
Mid-size	$32.95	$199.00
Full-Size	$39.95	$239.00
Van for 8	$45.95	$259.00

*150 free miles per day. Add $0.20 for each additional mile.

Norman Riley rented a mid-size car for 2 days and drove 358 miles. How many miles will he be charged for? How much was the rental cost?

Total miles		Free miles (150 × 2)		Chargeable miles
358	−	300	=	58

Basic charge		Additional charge		Rental cost
2 × $32.95		58 × $0.20		
$65.90	+	$11.60	=	$77.50

Complete each row of the table using prices from Carla's Car Rental. The first row is a sample based on the example above.

	Customer's Name	Type of Car	Length of Rental	Miles Driven	Rate (see above)	Basic Charge	Chargeable Miles	Additional Charge	Rental Cost
	N. Riley	Mid-size	2 days	358	$32.95	$65.90	58	$11.60	$77.50
1.	M. Wilson	Compact	2 days	195					
2.	R. Edwards	Mid-size	2 weeks	1295					
3.	B. Quigley	Van	5 days	1024					
4.	S. Schott	Van	3 days	764					
5.	S. Lee	Full-size	1 week	1500					

6. Milton Anderson wants to rent a compact car for 16 days. How many weeks and days is this? How much is the basic charge at Carla's Car Rental for that length of time? If 8% sales tax is added, how much is the total charge?

Answer ______________________

7. Liz Thompson does not own a car. She rents a compact car about 6 days per month, and drives less than 150 miles each time. How much does she spend each month for rental costs? Would she save money by buying a new car that would require payments of $250 per month?

Answer ______________________

Financing a Car: Monthly Payment

Once you have decided which car to buy and where to buy it, you need to decide where to borrow the money to pay for it. You can finance the car through the dealer or borrow money from a bank or finance company. Most banks do not lend the full price of a car. If you trade-in an older car, the dealer gives you a **trade-in allowance** that may serve as a down payment.

Usually a new car loan is for 3 years or more. Sometimes you can get a lower interest rate if you take out a loan for a shorter period of time. The table at the right can be used to find monthly payments. What is the monthly payment on a \$9200 loan for 4 years at 12%?

The payment is \$2.63 per \$100. So, the payment for a \$9200 loan is 92 × \$2.63 or \$241.96.

Monthly Payment per \$100			
Int. Rate	3 yr	4 yr	5 yr
11%	\$3.27	\$2.59	\$2.18
12%	\$3.32	\$2.63	\$2.22
13%	\$3.37	\$2.68	\$2.27
14%	\$3.42	\$2.73	\$2.33
15%	\$3.47	\$2.78	\$2.38
16%	\$3.52	\$2.83	\$2.43

Find the monthly payment. Use information from the table of monthly payments.

	Amount	Int. Rate	Years	Payment
1.	\$4400	13%	4	
2.	\$2000	14%	3	
3.	\$6200	11%	4	
4.	\$1800	14%	3	

	Amount	Int. Rate	Years	Payment
5.	\$3500	12%	3	
6.	\$7900	16%	5	
7.	\$5100	15%	5	
8.	\$8900	12%	4	

9. Sarah buys a car for \$11,385. She has no car to trade in. Her down payment is \$1385. If she takes out a loan at 13% interest for 5 years, how much will the monthly payments be?

Answer ______________

10. Morris buys a car for \$10,295. He gets a trade-in allowance of \$2295 for his old car. If he takes out a loan at 12% interest for 4 years, how much will the monthly payments be?

Answer ______________

11. Suppose Sarah decides to take out a loan at 12% interest for 3 years. How much will the monthly payments be? (See Exercise 9.)

Answer ______________

12. Suppose Morris gets \$500 more for the trade-in allowance. How much will the monthly payments be? (See Exercise 10.)

Answer ______________

Financing a Car: Total Interest Paid

The monthly payment on a 4-year loan of $9200 at 12% is $241.96. How much interest will be paid on the loan?

First find the number of months in 4 years.

$$4 \times 12 = 48$$

Then multiply the number of months by the amount paid each month, and subtract the amount borrowed.

$48 \times \$241.96 = \$11,614.08$
$\$11614.08 - \$9200 = \$2414.08$ Total interest

Monthly Payment per $100			
Int. Rate	3 yr	4 yr	5 yr
11%	$3.27	$2.59	$2.18
12%	$3.32	$2.63	$2.22
13%	$3.37	$2.68	$2.27
14%	$3.42	$2.73	$2.33
15%	$3.47	$2.78	$2.38
16%	$3.52	$2.83	$2.43

Complete each row of the chart. For rows 5 and 6, use the table to find the monthly payments.

	Amount of Loan	Interest Rate	Number of Years	Monthly Payment	Total of Payments	Total interest
1.	$10,000	13%	5	$227.00		
2.	$10,000	14%	5	$233.00		
3.	$8,000	15%	3	$277.60		
4.	$8,000	15%	5	$190.40		
5.	$4900	13%	4			
6.	$5500	11%	3			

7. Exercises 1 and 2 above differ only by the interest rate. How much less total interest will be paid if the interest rate is 13% instead of 14%?

Answer ____________________

8. Exercises 3 and 4 above differ only by the number of years. How much less total interest will be paid on the 3-year loan than the 5-year loan?

Answer ____________________

9. Morris buys a car for $14,295. He makes a down payment of $1295. If he takes out a loan at 12% interest for 4 years, how much will the monthly payments be?

Answer ____________________

10. Refer to Exercise 9. What will be the total of the monthly payments? What will be the total cost of the car, including interest?

Answer ____________________

Automobile Insurance: Base Premiums

Insurance is another large expense of operating a car. The cost is based on the sex and age of the driver, the size and value of the car, where the car is driven, and other factors. A car owner buys an **insurance policy** and pays **premiums** to an insurance company. The company then agrees to pay certain accident costs. Some policies have a **deductible clause.** A $100-deductible clause means that the insured person pays the first $100 of the repair bill, and the insurance company pays the rest.

The chart at the right shows the **base premiums** that one company charges for collision and comprehensive insurance. **Collision insurance** pays to repair the insured person's car after an accident. **Comprehensive insurance** pays to repair the car in case of fire, theft, or damage caused by storms.

YEARLY BASE PREMIUM				
Car Type	Collision		Comprehensive	
	$100Ded.	$250Ded.	$50Ded.	$100Ded.
A	$109.30	$79.40	$42.20	$31.40
B	120.50	111.60	62.10	47.30
C	134.70	162.50	78.30	59.90

What is the total yearly base premium for each of these policies? (Hint: Find the base premium for each amount of insurance and add.)

1. Car Type A, Collision with $250 deductible, Comprehensive with $50 deductible. ____________

2. Car Type C, Collision with $250 deductible, Comprehensive with $100 deductible ____________

3. Car Type B, Collision with $100 deductible, Comprehensive with $50 deductible. ____________

4. Car Type A, Collision with $100 deductible, Comprehensive with $100 deductible. ____________

In addition to collision and comprehensive insurance, many states require drivers to buy **liability insurance** of the following types. **Property damage insurance** pays to repair the damages to other people's cars or property. **Bodily injury insurance** pays for medical expenses after an accident.

The chart at the right shows annual base premiums charged by one company. A bodily injury limit of 50/100 means that the insurance company will pay up to $50,000 to each person injured, with a limit of $100,000 if more than one person is injured.

Yearly Base Premiums for Liability Insurance			
Bodily Injury Limits	Premium	Property Damage Limits	Premium
50/100	145.20	$25,000	$55.20
100/300	$160.80	$50,000	$59.30
250/500	$168.10	$100,000	$64.80

What is the total yearly base premium for each of these liability insurance policies? (Hint: Find the base premium for each amount of insurance and add.)

5. Bodily injury 100/300
Property damage $25,000 ____________

6. Bodily injury 50/100
property damage $100,000 ____________

7. Bodily injury 250/500
Property damage $50,000 ____________

8. Bodily injury 100/300
Property damage $50,000 ____________

Automobile Insurance: Driver-Rating Factor

After an insurance company finds the base premium for a policy, this amount is multiplied by the **driver-rating factor.** A person with a high driver-rating factor pays much more for insurance than a person with a low driver-rating factor. An insurance company may use a table like the one at the right to determine the driver-rating factor for an unmarried person. The factor would be lower for a married person and higher for someone who has had previous accidents.

DRIVER-RATING FACTORS FOR UNMARRIED PRINCIPAL OPERATORS		
MALES	**For Pleasure**	**For Work**
Under age 21	3.9	4.1
Age 21-24	2.9	3.1
Age 25-29	1.7	1.9
FEMALES	**For Pleasure**	**For Work**
Under age 21	1.8	2.0
Age 21-24	1.6	1.8
Age 25-29	1.0	1.2

Larry is an unmarried male, age 22, who drives for pleasure. What is Larry's driver-rating factor? How much will Larry pay for an insurance policy that has a base premium of $245 per year?

Larry's driver-rating factor is 2.9.
$2.9 \times \$245 = \$710.50$.
The policy will cost $710.50.

Use the table above to find the driver-rating factor for each person. Then find the total annual premium.

	Driver Description	Rating Factor	Annual Base Premium	Total Annual Premium
1.	25-year old unmarried female who drives for work		$183.90	
2.	19-year old unmarried male who drives for work		$207.70	
3.	18-year old unmarried female who drives for work		$152.40	
4.	28-year old unmarried male who drives for pleasure		$139.05	

Use the tables on page 110 to answer each question.

5. Marilyn Lee has a driver-rating factor of 2.0 and drives a type B car. What is her yearly premium for this policy?

- Bodily injury with limit of 50/100
- Property damage with limit of $50,000
- Collision with $250 deductible
- Comprehensive with $100 deductible

Answer ______________________

6. Jon Carlson has a driver-rating factor of 2.5 and drives a type C car. What is his yearly premium for this policy?

- Bodily injury with limit of 100/300
- Property damage with limit of $100,000
- Comprehensive with $50 deductible
- Collision with $100 deductible

Answer ______________________

Miles per Gallon

To find a car's gasoline mileage, divide the number of miles traveled by the number of gallons of gasoline used.

Jeff Sims kept a log book of his gasoline purchases. He always fills the tank. On January 31 he bought 16.2 gallons of gasoline for $22.52. How many miles did he travel between January 24 and January 31?

Date	Odometer	Gallons	Cost
1/24	42,891	Full	
1/31	43,293	16.2	$22.52
2/4	43,603	13.5	$19.58
2/10	44,010	15.8	$23.49
2/15	44,267	11.4	$16.85

43,293 − 42,891 = 402 miles

What was the gas mileage in miles per gallon (mpg)? Divide the number of miles traveled by the number of gallons used. Round the quotient to the nearest whole number.

402 miles ÷ 16.2 gallons = 24.81 or 25 mpg

What was the cost of gasoline per mile? Divide $22.52 by the number of miles driven. Round the quotient to three decimal places.

$22.52 ÷ 402 = $0.056 or 5.6¢ per mile

Complete each row of the table. Use data from the chart above.

	Date of Gasoline Purchase	Miles Traveled Since Last Purchase	Miles per Gallon	Cost per Mile (in cents)
1.	February 4			
2.	February 10			
3.	February 15			

4. Bart's car used 26 gallons of gasoline in 780 miles. How many miles per gallon was this?

Answer ____________

5. Joy's car used 12.6 gallons of gasoline in 288 miles. How many miles per gallon was this?

Answer ____________

6. Suppose gasoline costs $1.23 per gallon and you use 24 gallons per week. How much would you need to spend for gasoline each week?

Answer ____________

7. Teresa took a 540-mile trip. She used 19 gallons of gasoline and paid $1.28 per gallon. What was the cost per mile for gasoline?

Answer ____________

Mileage Table

If you have used a road map, you may have seen a mileage table like the one at the right. Before a trip is begun, the wise motorist will find out the distances so that she or he can make plans accordingly.

To find the mileage between two cities, find one of the cities in the left-hand column. Lay a ruler along that row to the column headed by the other city's name. Find the road distance from Atlanta to New York.

Atlanta is the first city in the left column. Move along that row to the column for New York. The distance is 868 miles.

AUTOMOBILE MILEAGE	Chicago, Ill.	Denver, Colo.	Houston, Tex.	Los Angeles, Calif.	Miami, Fla.	New York, N.Y.	San Francisco, Calif.
Atlanta, Ga.	671	1436	852	2245	663	868	2579
Boston, Mass.	992	2016	1865	3004	1615	220	3265
Chicago, Ill.		1062	1139	2115	1352	824	2240
Cleveland, Ohio	311	1393	1372	2393	1327	493	2571
Dallas, Tex.	923	820	241	1476	1367	1580	1790
Denver, Colo.	1062		1061	1148	2104	1794	1324
Detroit, Mich.	271	1333	1306	2415	1437	670	2511
Houston, Tex.	1139	1061		1585	1256	1714	1950
Las Vegas, Nev.	1905	843	1426	305	2572	2637	624
Los Angeles, Calif.	2115	1148	1585		2841	2784	411
Memphis, Tenn.	579	1054	560	1816	1050	1114	2214
Miami, Fla.	1352	2104	1256	2841		1395	3192
Montreal, Quebec	804	1866	1839	2948	1749	381	3044
New Orleans, La.	981	1326	360	1945	896	1360	2296
New York, N.Y.	824	1794	1714	2784	1395		3045
Philadelphia, Pa.	766	1770	1624	2726	1278	90	3021
Portland, Oreg.	2255	1329	2295	1098	3433	2992	687
St. Louis, Mo.	283	848	856	1832	1346	1057	2168
Seattle, Wash.	2076	1411	2377	1280	3421	2852	869
Toronto, Ontario	469	1531	1504	2613	1631	513	2709
Washington, D.C.	701	1696	1476	2680	1130	238	2968

1. How far is it from Atlanta to Los Angeles? ______________
2. How far is it from Boston to San Francisco? ______________
3. From Chicago to Los Angeles is how many miles by road? ______________
4. From Detroit to New York is how far? ______________
5. How far is it from Denver to Philadelphia? (It does not make any difference which one of the cities is in the left-hand column and which is at the top. The distance between the two cities is the same, in either direction.) ______________
6. Margaret and her family took a trip by car from Miami to Memphis. How many miles is this? ______________
7. The average speed for the trip was 50 miles per hour. How many driving hours did it take to make the trip? ______________
8. They averaged 28 miles to the gallon of gasoline. How many gallons did they use? ______________
9. At an average price of $1.48 a gallon, how much did the gas for the trip cost? ______________
10. The Fontaine family drove from Montreal to Miami. How many miles is this? ______________
11. They averaged 30 miles to the gallon of gas on the trip. How many gallons did they use? ______________
12. At $1.53 a gallon on the average, how much did the gas cost? ______________
13. How far is it from Houston to Las Vegas? ______________

Work Backwards

Sometimes you are told what happened at the end of a problem and asked to find what happened at the beginning. You can find the solution to a problem like this by working backwards.

Read the problem.

The final sales price of a new car was $9849. This price included sticker price, $555 tax, and a $44 license fee. What was the sticker price?

Use the strategy.

Sticker price	+	Tax	+	License fee	=	Final price
???	+	$555	+	$44	=	$9849

To find the sticker price, work backwards from the final price of $9849. Undo each operation by using the inverse operation. For example, you would use subtraction to undo addition. (You would use division to undo multiplication.)

$9849 – $44 – $555 = Sticker price

First subtract $44. $9849 – $44 = $9805
Then subtract $555. $9805 – $555 = $9250

The sticker price was $9250.

Solve by working backwards.

1. A used car dealer agreed to reduce the price of a used car to $2300. This was $599 less than the dealer's original price. How much was the original price?

Answer ____________________

2. A customer ordered a car that had a sticker price of $13,721. This included the base price of $11,939, destination charge of $309, and several options. What was the total cost of the options?

Answer ____________________

3. A customer ordered a car that had a sticker price of $8350. The sticker price included options totaling $1285 and a destination charge of $249. What was the base price?

Answer ____________________

4. Marla rented a car and was charged a total of $63.96. This included a basic charge of $49.00 plus an additional charge for 68 miles. How much was she charged for each additional mile?

Answer ____________________

Solve.

5. Bradley rented a car and was charged a total of \$105.05. He paid for 3 days at the daily rate plus \$6.20 for additional miles. What was the daily rate for renting the car?

Answer ______________

6. Greg borrowed for a car and is making payments of \$147.15 per month for 36 months. When he finishes paying off the loan, he will have paid a total of \$797.40 in interest. What was the amount of the loan?

Answer ______________

7. The total base premium for Eugene's two types of liability insurance is \$215.30. The base premium for bodily injury insurance is \$131.90. What is the base premium for the property damage insurance?

Answer ______________

8. Yolanda Everett has a driver-rating factor of 1.8. Her total annual premium for car insurance is \$388.80. What is the annual base premium?

Answer ______________

9. When Paula Pearson bought gasoline on March 8, the odometer reading was 34,901. This was 329 more than the odometer reading on March 1. What was the odometer reading on March 1?

Answer ______________

10. Oscar bought a road map and 12.5 gallons of gasoline for \$20.50. The road map cost \$2.00. What was the price per gallon of gasoline?

Answer ______________

DRIVING A CAR

Calculator Applications

Some problems require more than one operation. Suppose you use a calculator to find the value of $549 - 36 \times 8$, as shown at the right.

Standard Calculator

549 [−] 36 [×] 8 [=] 4104.

Why are there two different answers? A standard calculator does the operations one at a time as they are entered. A scientific calculator does multiplication and division before addition and subtraction. Which type of calculator do you have?

Scientific Calculator

Since $549 - 36 \times 8$ should have only one value, parentheses are used to show which operation to do first. Which calculator found the value of $(549 - 36) \times 8$? Which calculator found the value of $549 - (36 \times 8)$?

If you are using a scientific calculator and you want it to complete an operation, press [=] before entering the next operation.

EXAMPLE When Maxwell filled the gas tank of his car, the odometer reading was 14,285. When he filled the tank previously, the reading was 13,977. He bought 10.4 gallons of gas. What is the car's gas mileage?

Find the value of $(14{,}285 - 13{,}977) \div 10.4$

14285 [−] 13977 [=] [÷] 10.4 [=] 29.615384

The gas mileage is 29.6 miles per gallon.

Use a calculator to find each value. Round to the nearest tenth.

1. $984.2 - (845 \div 5)$ ____________

2. $(984.2 - 845) \div 5$ ____________

3. $1000 \div (14.6 + 8.4)$ ____________

4. $(1000 \div 14.6) + 8.4$ ____________

5. \$8392 − (\$530 + \$239) ____________

6. (\$8392 − \$530) + \$539 ____________

Solve.

7. When Jenny filled the gas tank of her car, the odometer reading was 58,513. When she filled the tank previously, the reading was 58,216. She bought 12.3 gallons of gas. What is the car's gas mileage?

Answer ____________

8. When Dave filled the gas tank of his car, the odometer reading was 25,542. He bought 9.7 gallons of gas and figured the gas mileage was about 33 miles per gallon. What was the odometer reading when he filled the car with gas previously?

Answer ____________

Depreciation

The value of a car goes down every year. This decrease in value is called **depreciation.** A new car depreciates most during the first year. Then it depreciates at a slower rate. Some cars depreciate more quickly than others. Guidebooks that list used car values can be helpful in finding out how much a car has depreciated, and, therefore, how much it is worth at the time.

Depreciation is often given as a percent of the purchase price. The approximate trade-in or sale value of a car can be found by subtracting the amount of depreciation from the car's purchase price.

The purchase price of a new car was $8500. In the first year, its value depreciated 30%. In the second year, its value depreciated 15%. Find the amount of depreciation each year and the approximate trade-in or sale value after two years.

First year depreciation	Second year depreciation	Approximate trade-in or sale value after 2 years
30% of $8500	15% of $8500	
0.30 × $8500 = $2550	0.15 × $8500 = $1275	$8500 − $2550 − $1275 = $4675

The approximate trade-in or sale value after two years is $4675.

Find the amount of depreciation each year, for the given rates. Then find the approximate trade-in value after 4 years.

	Purchase Price of Car	Year 1 30%	Year 2 15%	Year 3 12%	Year 4 8%	Approximate Trade-In Value
1.	$10,000	0.30 × 10,000				
2.	$12,430					
3.	$9820					
4.	$13,800					
5.	$6740					

6. If the value of a car depreciates $3000 in one year, what is the average depreciation per month? (Hint: Divide the amount by the number of months in a year.)

Answer ____________________

7. The trade-in value of Ruth Gorman's car is $3500. She purchased the car for $8180 and has had it for 36 months. What was the average amount of depreciation per month?

Answer ____________________

DRIVING A CAR

Car Maintenance

Another expense of operating a car is maintenance to keep the car in good running condition. Some items included in car maintenance are oil, oil filters, grease, fuel filters, spark plugs, and tires. Many people have their cars maintained by service stations, shops, or car dealers.

Ria Waterson had her car serviced at Red's Service Station. The **invoice,** or bill, at the right shows what work was done, how many hours of labor are being charged, and how much the parts cost. Study the amounts on the invoice. This is how to compute the total charge.

First, add to find the total number of hours of labor.

0.25 + 0.5 + 0.25 + 0.5 = 1.5 hours

Add to find the total cost of the parts.

$6.45 + $16.00 + $7.35 + $3.85 = $33.65

From the invoice you can see that the labor charge was $43.00 per hour. Multiply the hours by $43 to find the labor charge.

1.5 hours × $43/hour = $64.50

Add the labor charge to the parts charge.

$33.65 + $64.50 = $98.15

RED'S SERVICE STATION		
Customer: name: *Ria Waterson*		
Auto license number: *MDY 330*		
Work done	**Hours**	**Parts**
Replace:		
Oil Filter	*0.25*	*$6.45*
Spark Plugs (set of 4)	*0.5*	*16.00*
Air Filter	*0.25*	*7.35*
Fuel Filter	*0.5*	*3.85*
Totals	*1.5*	*$33.65*
Labor charge: Number of Hours × 43.00 =		*$64.50*
Total Charge: Parts + Labor =		*$98.15*

Solve.

1. Complete the invoice form below to find the total charge.

PAT'S AUTO SERVICE		
Customer: *Henry Ruscilli*		
Auto license number: *GXR 498*		
Work done	**Hours**	**Parts**
Rotate Tires	*1.0*	*0*
Spark Plugs (set of 6)	*0.75*	*19.50*
Air Filter	*0.25*	*6.95*
Fuel Filter	*0.25*	*4.50*
Totals		
Labor charge: Number of Hours × 45.00 =		
Total Charge: Parts + Labor =		

2. A different customer at Red's Service Station above was charged for 3.25 total hours of labor. What was the labor charge?

Answer ____________________

3. A service station charges a total of $20.60 to change the oil and oil filter on Nathan's car. Nathan can buy an oil filter for $2.50 and the oil for $4.20 at a discount store. How much will he save if he does the work himself?

Answer ____________________

4. Vera took her car to a service station that charged $45.00 per hour for labor. If the invoice included $47 for parts and 2.5 hours for labor, what was the total charge?

Answer ____________________

Estimates for Car Repairs

Because cars are expensive to maintain and repair, a car owner needs to allow money in the budget to cover these costs. Also, a car owner should search for a reliable repair shop that has reasonable prices. One way to find a reliable shop is to ask friends and neighbors for references.

When major repairs are needed, the owner of a car may want to get estimates from two or three different shops. Three estimates for the same work are shown below.

	ESTIMATES		
Repair	**Shop A**	**Shop B**	**Shop C**
Replace brakes	$173.78	$251.00	$178.50
Align front end	40.00	36.88	38.45
Service air conditioning	19.88	23.35	21.66
Service transmission	52.50	46.78	48.50
Tune up engine (6 cyl)	49.65	38.44	39.88
TOTALS:			

1. Complete the table to find the total for the repairs at each shop.

2. Which shop has the lowest estimate for these repairs? How much less than the most expensive shop does it charge?

Answer ______________________

3. Which shop charges the least for replacing brakes? How much less than the most expensive shop does it charge?

Answer ______________________

4. How much would you save by having the transmission serviced at Shop B rather than at Shop A?

Answer ______________________

5. Suppose you want to have the brakes replaced and the engine tuned at the same shop. From which shop would you get the best price? How much would you pay at that shop?

Answer ______________________

6. Suppose you want to have the front end aligned and the transmission serviced at the same shop. What is the best price you could get?

Answer ______________________

7. Paul's car was in an accident and needed a new fender. Paul got three different estimates for repairs: $790, $912, and $944. Paul has a $200-deductible clause in his insurance coverage. The insurance company paid the lowest estimate, except for the deductible. How much was paid by the insurance company?

Answer ______________________

Public Transportation

Most cities have various kinds of public transportation, such as buses and trains. The companies that operate the public transportation often sell monthly passes in addition to single-ride tickets. If a person uses public transportation frequently, buying a monthly pass may be a way to save money. Riding buses or trains is usually less expensive than owning and operating a car.

Teresa Porter rides the train to work and back 22 days in a month. A single-ride ticket costs $0.90 and a monthly pass costs $28.50. How much money would she save by buying a monthly pass instead of single-ride tickets?

Single-ride ticket		Days per month		Rides per day		Total cost
$0.90	×	22	×	2	=	$39.60

Find the difference between $39.60 and $28.50.
She would save $11.10 each month.

Solve.

1. Suppose you ride a bus to and from work 24 days per month. A single-ride ticket costs $1.20. A monthly ticket costs $42.00. Which ticket will cost less to use each month? How much less?

Answer ____________________

2. Use the information in Exercise 1. Suppose you took a 5-day vacation in one month. Would you pay more or less if you bought a monthly ticket for that month? How much more or less?

Answer ____________________

3. Suppose you ride a train to and from work 20 days per month. A single-ride ticket costs $3.75. A monthly ticket costs $105.00. How much will you save each month by using a monthly pass?

Answer ____________________

4. Vernon drives to work some days and rides a bus other days. The bus costs $1.65 each way. When he drives, he pays $4.00 per day for parking and about $2.20 for gasoline. How much less does it cost each day to ride the bus?

Answer ____________________

5. Here is a fare schedule for a bus company.

	Monday-Friday	Saturday-Sunday
Adults	$1.50	$1.00
Children	$1.00	$0.75

How much will a family of 2 adults and 3 children pay for a round trip on a Monday?

Answer ____________________

6. Use information from Exercise 5. How much will the same family pay if they ride on a Saturday? How much less will they pay on Saturday than on Monday?

Answer ____________________

Airline Schedules

The table at the right is a sample airline schedule. It gives hours of departure and arrival, flight numbers, and number of stops. Note that the times are either central daylight time (CDT) or eastern daylight time (EDT).

Reggie Smolen has a ticket on flight 329 from Kansas City to Minneapolis. How much time is required for the flight?

Find flight 329 in the Kansas City to Minneapolis section of the table. Read across. The flight leaves at 7:45 P.M. and arrives at 9:40 P.M. The flight takes 1 hour and 55 minutes.

	Leave	Arrive	Flight	Stops
CHICAGO (CDT) to HOUSTON (CDT)	8:50 A.M.	2:47 P.M.	239	4-Stop
	2:40 P.M.	4:53 P.M.	531	Non-Stop
	7:15 P.M.	10:47 P.M.	68	1-Stop
	9:00 P.M.	12:22 A.M.	374	1-Stop
	11:20 P.M.	1:33 A.M.	286	Non-Stop
KANSAS CITY (CDT) TO MINNEAPOLIS (CDT)	7:25 A.M.	9:15 A.M.	169	1-Stop
	9:50 A.M.	10:56 A.M.	184	Non-Stop
	11:25 A.M.	1:20 P.M.	246	1-Stop
	2:15 P.M.	4:10 P.M.	381	1-Stop
	5:05 P.M.	6:11 P.M.	62	Non-Stop
	6:00 P.M.	7:55 P.M.	418	1-Stop
	7:45 P.M.	9:40 P.M.	329	1-Stop
	9:10 P.M.	10:16 P.M.	172	Non-Stop
	9:15 P.M.	10:59 P.M.	96	1-Stop
WASHINGTON (EDT) to MIAMI (EDT)	7:00 A.M.	11:30 A.M.	218	3-Stop
	8:05 A.M.	11:03 A.M.	8	1-Stop
	9:45 A.M.	12:01 P.M.	36	Non-Stop
	2:00 P.M.	4:13 P.M.	472	Non-Stop
	4:30 P.M.	6:45 P.M.	10	Non-Stop
	6:45 P.M.	8:57 P.M.	302	Non-Stop
	9:30 P.M.	12:43 A.M.	181	1-Stop

Note: Departing and arriving times are based on time zones in each city.

Use the airline schedule to answer each question.

1. Flight 329 makes 1 stop before arriving in Minneapolis. Look at flight 172. How much less time does the nonstop flight take?

Answer ________________

2. Mr. Bolten lives in Chicago. He has a ticket on the first afternoon plane to Houston. What time does the plane leave?

Answer ________________

3. What time will it be in Houston when Mr. Bolten arrives?

Answer ________________

4. How much time is required for flight 218 from Washington to Miami?

Answer ________________

5. If you take flight 36 instead of flight 218, how long will the trip take?

Answer ________________

6. How many stops does flight 239 from Chicago make before landing in Houston?

Answer ________________

7. Mrs. Templeton wants to fly from Kansas City to Minneapolis. She needs to arrive before 7 P.M. What is the flight number of the latest afternoon plane she can take?

Answer ________________

8. When will Mrs. Templeton arrive in Minneapolis?

Answer ________________

9. How much time is required for flight 62 from Kansas City to Minneapolis?

Answer ________________

10. How much more time does flight 418 take than flight 62?

Answer ________________

Find a Pattern

By studying patterns, you may be able to solve problems or discover mistakes that have been made. When one number in a problem is changed, the result may change according to a certain pattern.

Read the problem.

Three different customers rented compact cars from the same rental company. This chart shows the length of time the cars were rented and the rental cost. If a compact car is rented for 9 days, what could be the rental cost?

Customer	Number of Days	Rental Cost
A	7	$169.00
B	10	$241.00
C	8	$193.00

Use the strategy.

Study the differences between the costs. Customer A rented a car for the least number of days and paid the least amount. Customer C rented a car for one day more than Customer A, and paid $193 − $169 or $24 more. Customer B rented a car for two days more than Customer C, and paid $241 − $193 or $48 more.

The pattern seems to be $24 more for each additional day. So a 9-day rental would probably cost $24 more than an 8-day rental.

$193 + $24 = $217

The rental cost for 9 days could be $217.

Solve.

1. At a car rental agency, one customer was charged $14.40 for 60 miles. A different customer was charged $19.20 for 80 miles. How much might a customer pay for 70 miles?

Answer ________________

2. Three car buyers got 4-year loans at the same interest rate. Person A borrowed $10,000 and pays $259 per month. Person B borrowed $11,000 and pays $284.90 per month. If Person C borrowed $12,000, how much is the monthly payment?

Answer ________________

Solve.

3. Fred made a mistake when he copied the monthly payment amounts for a 3-year loan of $6000.

Rate	12%	13%	14%	15%
Payment	$179	$202	$205	$208

Which amount is wrong? What might the correct amount be?

Answer ____________________

4. A customer wants to borrow $8500 for 3 years. The payments on an $8000 loan would be $273.60. The payments on a $9000 loan would be $307.80. How much will the payments be for the loan of $8500?

Answer ____________________

5. A person with a driver-rating factor of 2.0 pays $240 for a certain policy. The same policy costs $300 for a person with a driver-rating factor of 2.5. How much would the same policy cost for a person with a driver-rating factor of 3.0?

Answer ____________________

6. Jo Ellen made a list of her gasoline purchases for one month. She paid the same price per gallon each time. Jo Ellen made a mistake on one of the prices. Which price is probably wrong? What might the correct price be?

10 gal for $13.99   14 gal for $19.60
15 gal for $29.90   11 gal for $15.39

Answer ____________________

7. Several cars were driven the same distance but used different amounts of gasoline, as shown in the list below. For some cars, the rate of gas usage in miles per gallon (mpg) is also listed. What are the missing numbers in the list?

Car	Gallons	mpg
A	20	50
B	25	40
C	40	25
D	100	10
E	50	______
F	80	______

8. Here is a list of labor costs at a repair shop, but there is a mistake in one of the amounts. Cross out the mistake and write the correct amount.

8 hours	$232.00
11.5 hours	$353.50
5 hours	$145.00
3.5 hours	$101.50

Answer ____________________

Yearly Automobile Expenses

Many people do not realize how much they spend to own and operate a car. Some of the costs, such as maintenance and gasoline, vary from month to month. Other costs, such as insurance and license plate fees, may stay the same each year.

The charts show estimates of two persons' automobile expenses. Find the yearly and monthly expenses in each category. Then add to find the total yearly and monthly expenses.

	Category	Jack's Estimated Expenses	Yearly Expenses	Monthly Expenses
1.	Depreciation (second year)	15% of $9500		
2.	Insurance	$324 twice per year		
3.	Gasoline	$15/wk × 52 weeks (average)		
4.	Maintenance	$180 twice per year		
5.	License plate fees	$65 per year		
6.	TOTAL			

	Category	Jill's Estimated Expenses	Yearly Expenses	Monthly Expenses
7.	Depreciation (first year)	30% of $12,000		
8.	Insurance	$285 twice per year		
9.	Gasoline	$24/wk × 52 weeks (average)		
10.	Maintenance	$120 twice per year		
11.	License plate fees	$75 per year		
12.	TOTAL			

13. If Jack drives 11,000 miles this year, about how much is the cost per mile? (Use the answer from Exercise 6.)

Answer ____________________

14. If Jill drives 13,000 miles this year, about how much is the cost per mile? (Use the answer from Exercise 12.)

Answer ____________________

Unit 5 Review

Solve.

1. Find the sticker price of a new car that has a base price of $8624, destination charge of $165, and these options:

Air conditioning	$597
Rear-window defroster	$89

Answer ____________

2. Maria rented a minivan for $54 per day for 8 days and drove 1452 miles. She was allowed 150 free miles per day, and was charged $0.25 for each additional mile. What was the total rental cost?

Answer ____________

3. Cory borrowed $9000 and must make payments of $303 per month for 3 years. What is the total of the monthly payments? What is the total amount of interest paid?

Answer ____________

4. The yearly base premium for Wally's automobile insurance is $197.60. If his driver-rating factor is 3.7, how much is his yearly premium?

Answer ____________

5. Rosa filled her car's gasoline tank on April 6. Find the number of miles she has driven since April 1, and the miles per gallon.

Date	Odometer	Amount
April 1	62,458	full
April 6	62,702	11.4 gal

Answer ____________

6. Suppose gasoline costs $1.15 per gallon. If your car gets 30 miles per gallon, find the cost of gasoline for a 900-mile trip.

Answer ____________

7. A car was purchased for $12,300. Its value depreciated 65% in the first four years. Find the approximate trade-in value after four years.

Answer ____________

8. Kyle had his car's air conditioning serviced. The repair shop replaced 4 pounds of refrigerant and charged $9.65 per pound. The labor charge was $22.85. What was the total charge?

Answer ____________

9. Laura can buy a monthly train pass for $68 or single-ride tickets for $2.30. She rides the train to and from work 16 days per month. How much money would she save each month by buying a monthly pass?

Answer ____________

10. Here are Zachary's approximate car expenses for one year. Find the total yearly expenses.

Insurance	$720
Depreciation	$1500
License	$32
Gasoline	$960
Maintenance	$500

Answer ____________

Buying a House: Down Payment

Few people can afford to pay cash when purchasing a house. Instead, they make a cash down payment and borrow the rest of the money. The amount of the down payment is usually between 10% and 40% of the selling price of the house. The money borrowed is called a **mortgage** loan. There are some special funds set aside for qualified first-time buyers that have lower interest rates and require less of a down payment.

EXAMPLE 1 Alice and George Ramirez are buying a house for $62,500. A 15% down payment is required. What is the amount of the mortgage loan?

First, find the down payment: 0.15 × $62,500 = $9375

Selling price		Down payment		Mortgage loan
$62,500	−	$9375	=	$53,125

EXAMPLE 2 Mary Greer bought a $57,000 house and made a $9000 down payment. She must pay the balance in $576 monthly payments which include 12% interest. How much of the first month's payment is interest? How much is applied to the balance due?

The amount of the mortgage loan is $57,000 − $9000 or $48,000. Find the amount of the first month's payment that will be interest. Subtract to find the amount that will apply to the principal.

Principal		Rate		Time		Interest
$48,000	×	0.12	×	$\frac{1}{12}$	=	$480

Payment		Interest		Principal
$576	−	$480	=	$96

Complete the table to find the down payments and mortgage loans.

	Buyer	Selling Price	% Down	Down Payment	Mortgage Loan
1.	Linda Pinchfield	$80,500	20%		
2.	Joe and Sue Wong	$120,000	25%		
3.	Arnie Thompson	$90,000	15%		
4.	Ed and Tom Brown	$110,000	35%		

5. If Mary Greer had compared mortgage rates at several banks, she could have obtained a loan at 10% interest. How much would she have owed for interest the first month?
Refer to Example 2.

Answer ______________

6. Ron Genicola heard about a special fund to help first-time buyers. He bought a house for $60,000. He could make a 5% down payment and borrow the remaining amount at 8% interest for 30 years. His monthly payment is $418. How much of the first payment is interest?

Answer ______________

Buying a House: Closing Costs

A number of fees are included in the cost of buying a house. These fees, called **closing costs,** may include fees for the bank's lawyers, credit checks, property taxes, survey and inspection of property, and preparation of documents. Financing charges may also include **points**. One point is 1% of the loan. To find the closing costs, add all the bank fees.

1. Eleanor Phillips is getting a $40,000 bank loan. Complete the table to find her closing costs.

Fee	Amount
Credit report	$ 35
Loan origination: 2% of loan	
Abstract of title	60
Attorney fee	100
Document stamp: 0.3% of loan	
Processing fee: 1% of loan	
Points: 1% of loan	
TOTAL	

2. John Reilly is buying a $72,800 home with a $60,000 mortgage loan. Find his total closing costs.

Fee	Amount
Appraisal fee	$ 35
Credit report	25
Title search	175
Legal fees	250
Service fee: 2% of loan	
Processing fee: 1% of loan	
Points: 2% of loan	
TOTAL	

3. Jack Mendez is getting a $50,000 mortgage loan. Find his closing costs.

Fee	Amount
Credit report	$ 40
Loan origination: 2% of loan	
Attorney fee	300
Processing fee: 1% of loan	
Title and insurance	45
Points: 1.5% of loan	
TOTAL	

4. Sharon Urich is getting a $45,000 condominium with a $35,000 mortgage loan. Find her closing costs.

Fee	Amount
Credit report	$ 25
Title search	85
Service fee: 1.5% of loan	
Legal fees	150
Processing fee: 1% of loan	
Geologic survey	110
Points: 3% of loan	
TOTAL	

5. Refer to Exercise 2. John has saved $14,800. How much is John Reilly's down payment and closing costs? Does he have enough money saved?

Answer ____________________

6. Refer to Exercise 4. Sharon has saved $13,000. How much is Sharon's down payment and closing costs? Does she have enough saved?

Answer ____________________

Mortgages: Monthly Payments

Banks charge interest when giving mortgage loans. A table such as the one at the right can be used to find monthly mortgage payments for different interest rates and time periods. Each mortgage payment includes interest and an amount that is applied to reduce the loan.

Monthly Payments per $1000		
Interest Rate	15-Year Loan	30-Year Loan
6.0%	$8.44	$6.00
6.5%	$8.71	$6.32
7.0%	$8.99	$6.65
7.5%	$9.27	$6.99
8.0%	$9.56	$7.34
8.5%	$9.85	$7.69
9.0%	$10.15	$8.05
9.5%	$10.45	$8.41
10.0%	$10.75	$8.78

EXAMPLE Jake Dempsey borrowed $80,000 at an interest rate of 7.5% for 30 years. What is his monthly mortgage payment?

The table shows the payment is $6.99 per $1000. First find the amount of the loan in thousands.

$$\$80{,}000 \div \$1000 = 80$$

Then multiply by $6.99.

$$80 \times \$6.99 = \$559.20$$

Each monthly payment is $559.20.

Solve.

1. Joyce Rogers has a mortgage loan for $120,000 at 6% for 30 years. What is her monthly mortgage payment?

Answer ____________________

2. Leon and Cora Winters have a mortgage loan of $70,000 at 7.5% for 15 years. How much is their monthly payment? What would it be if they had the loan for 30 years?

Answer ____________________

3. The Kelsey family has a mortgage loan of $90,000 for 30 years at 7%. If they make every monthly payment for the 30-year period, how much will they pay in all? [Hint: Multiply the monthly payment by the number of months in 30 years.]

Answer ____________________

4. How much interest did the Kelsey family in Problem 3 pay for their mortgage loan? [Hint: Subtract the amount they borrowed from the total amount they paid back over the 30-year period.]

Answer ____________________

5. If the Kelsey family in Problem 3 decides to have a mortgage for 15 years at 7% instead of for 30 years at 7%, how much will they pay in all? How much interest will they save?

Answer ____________________

6. The Santos family wants to borrow $120,000. One lender offers them a 30-year mortgage at 8%. Another lender offers them the same loan, but at 7.5%. If they accept the second offer, how much will they save over the 30 years?

Answer ____________________

Mortgages: Adjustable Rate

Interest rates can vary from year to year. Because of this, most banks offer two kinds of mortgages. One is a **fixed-rate mortgage.** This kind has the same interest rate for each and every year of the loan. The second kind is an **adjustable-rate mortgage.**

An adjustable-rate mortgage is usually a 30-year loan with an interest rate that rises or falls according to the rate that banks must pay to borrow money. Because such loans carry less risk for the bank, they often carry a lower interest rate than the fixed-rate mortgages. Adjustable-rate mortgages often include a limit on the amount the rate can change during a given period, such as 2% per year.

Monthly Payments per $1000		
Interest Rate	15-Year Loan	30-Year Loan
6.0%	$8.44	$6.00
6.5%	$8.71	$6.32
7.0%	$8.99	$6.65
7.5%	$9.27	$6.99
8.0%	$9.56	$7.34
8.5%	$9.85	$7.69
9.0%	$10.15	$8.05
9.5%	$10.45	$8.41
10.0%	$10.75	$8.78

Complete the table to compare the monthly payments for the first year.

	Amount of Loan	30-Year Fixed-rate	Monthly Payment	30-Year Adjustable	Monthly Payment	Monthly Difference
1.	$50,000	7.0%	*$332.50*	6.0%	*$300.00*	*$32.50*
2.	$80,000	8.0%		6.5%		
3.	$100,000	9.0%		7.5%		
4.	$120,000	8.5%		7.0%		
5.	$60,000	7.5%		6.0%		

Solve.

6. The Wisneski family took out a 30-year adjustable mortgage for $80,000 at 6%. If the interest rate for a fixed-rate mortgage is 8%, how much less is their monthly payment?

Answer ____________________

7. How much less does the Wisneski family pay for their mortgage during the first year?

Answer ____________________

8. At the end of the first year, the interest rate on the Wisneski family's mortgage went up to 7%. What was their new monthly payment? (Use $80,000 as the amount of the loan.)

Answer ____________________

9. At the end of the second year, the interest rate on the Wisneski family's mortgage went up to 8.5%. Find how much more they pay per year for an adjustable-rate mortgage than a fixed-rate mortgage of 8%. (Use $80,000 as the amount of the loan.)

Answer ____________________

OWNING A HOME

Using Electricity

Electrical power is measured in watts or kilowatts. A **kilowatt** is equal to 1000 watts, or the amount of electricity used by ten 100-watt light bulbs. Electrical usage is measured in **kilowatt-hours.** If you shine ten 100-watt bulbs for 1 hour, you use one kilowatt-hour of electricity.

If you know the number of watts that an appliance uses, you can estimate the electrical usage for a length of time.

EXAMPLE Suppose a 600-watt microwave oven is used about 15 hours per month. How many kilowatt-hours does the oven use per month? If the electric company charges 8.2 cents per KWH, what is the cost of electricity for the microwave oven for a month?

Watts				Kilowatts
600	÷	1000	=	0.6

Kilowatts		Hours		KWH
0.6	×	15	=	9 KWH

The microwave oven uses 9 KWH if it is used 15 hours per month. The cost is 8.2 cents × 9, or 73.8 cents per month.

Solve.

1. An electric refrigerator uses 250 watts of electricity. How many kilowatt-hours will it use in 30 days if the refrigerator runs 9 hours per day?

Answer ____________________

2. Suppose six 60-watt light bulbs are turned on for 3 hours each day. How many kilowatt-hours of electricity are used by those bulbs in seven days?

Answer ____________________

The electrical usage of appliances varies depending on size, the efficiency of the appliance, and the operating conditions. For example, a refrigerator uses more power if the door is opened frequently.

The chart at the right shows the average number of kilowatt-hours that various appliances use each month. Find the average cost of operating each appliance if the electricity costs 8 cents per KWH.

	Appliance	Usage per Month (KWH)	Monthly Cost
3.	Clock	2	
4.	Clothes dryer	83	
5.	Water heater	400	
6.	Vacuum cleaner	4	
7.	Washing machine	37	
8.	Color television	46	
9.	Range with oven	98	
10.	Stereo	109	
11.	Iron	12	
12.	Toaster	3	

Reading Electric & Gas Meters

Most homes have electric meters with four dials like the ones shown below. These dials move to show the usage of electricity in kilowatt-hours. Usually once a month, someone from the electric company reads the meter. By subtracting the meter reading from the previous month's readings, the company can tell how much electricity you used in one month.

To read the meter, begin at the leftmost dial. Record the smaller of the two numbers that the hand on the dial is between. Do this for all the dials. The four numbers show the current reading of kilowatt-hours (KWH) used.

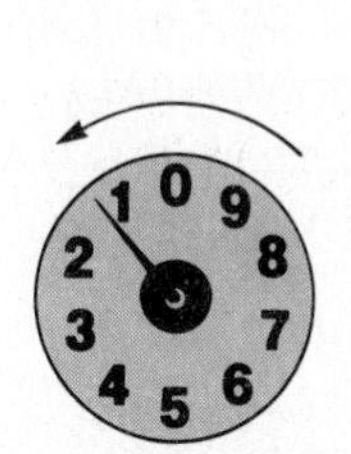

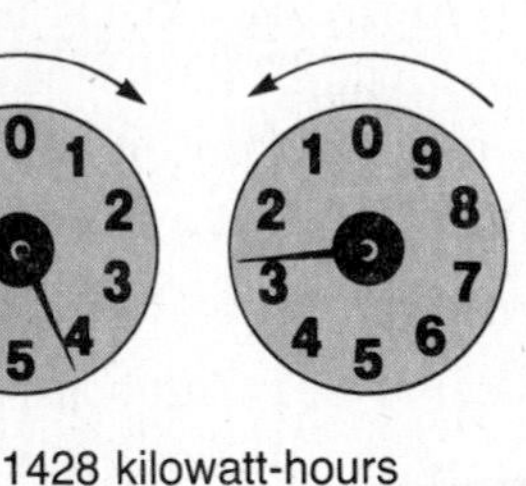

1428 kilowatt-hours

EXAMPLE Suppose the meter above shows a customer's reading for August 1. If the reading for July 1 was 0637, how many kilowatt-hours were used during July? If the cost of electricity is $0.07 per KWH, how much will the customer be charged?

Subtract to find the number of KWH used.	1428 − 0637 = 791
Multiply 791 by the cost per KWH.	791 × $0.07 = $55.37

Gas meters have three dials that are read in the same way as electric meters. Gas usage is measured in units of a hundred cubic feet, or CCF.

Write the meter reading in KWH's (kilowatt-hours).

1. 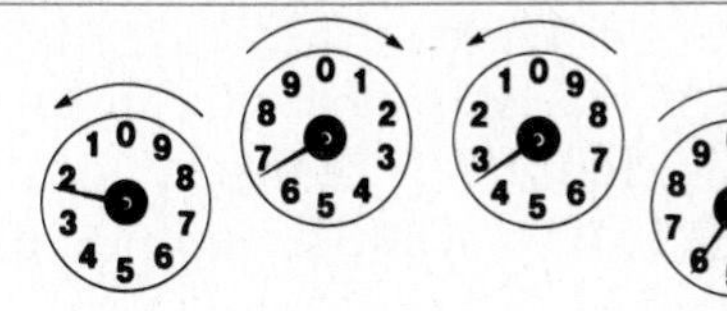

Answer ____________________

2. 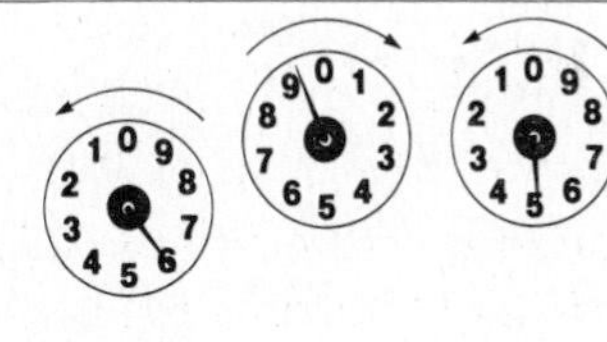

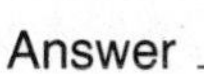

Answer ____________________

Write the meter reading in CCF's (hundred cubic feet).

3. 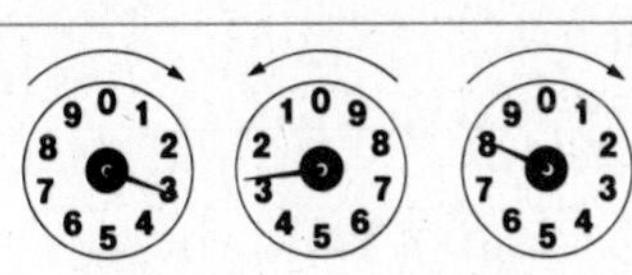

Answer ____________________

4. 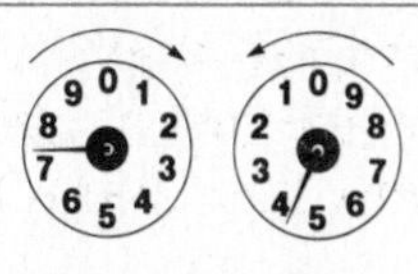

Answer ____________________

5. Last month Bertha's electric meter read 1574 KWH. This month it reads 1838 KWH. How many kilowatt-hours did she use during the month? How much is the bill if the cost is $0.08 per KWH?

Answer ____________________

6. Last month Ross Taylor's gas meter read 340 CCF. This month the reading was 496 CCF. He pays $0.52 per CCF, plus a service charge of $2.50. What was his bill?

Answer ____________________

Property Insurance

To get a mortgage loan from a bank, you must buy insurance for your house or property. Insurance can financially protect you or the mortgage holder from loss by fire or theft. Many people who rent also have insurance to protect their belongings.

EXAMPLE Al Wong's basic homeowner's insurance for his house and contents costs $240 per year. This cost is called an **insurance premium.** In addition to the basic insurance, Al wants to insure a camera that costs $550. If the company charges $0.90 per $100 of value, what will Al's total insurance premium be?

Value of camera in hundreds: $550 ÷ $100 = 5.5
Cost of insuring the camera: 5.5 × $0.90 = $4.95
Total insurance cost: $240 + $4.95 = $244.95

Complete this table to find the annual insurance premium for each item.

	Item	Value	Rate per $100 Value	Annual Premium
1.	Gold ring	$800	$1.70	
2.	Coin collection	$700	$2.50	
3.	Electric guitar	$1200	$3.00	
4.	Stamp collection	$2000	$1.50	

5. Marge Wiley's homeowner's insurance has a basic premium of $340. She also wants to insure a violin worth $400. The rate for musical instruments is $2.50 per $100 of value. What is her total premium?

Answer ____________________

6. Paul and Shirley Hartwig want to insure a condominium for $60,000. The insurance rate is $6.00 per $1000. What will their annual premium be?

Answer ____________________

7. Peter Wolcott built a frame house that he wants to insure for $80,000. The rate is $6.60 per $1000. What is his premium?

Answer ____________________

8. If Peter puts a tile roof on his house, the insurance rate will go down to $6.50 per $1000 because tile is fireproof. How much will he save per year on insurance costs if he adds a tile roof?

Answer ____________________

OWNING A HOME

Rental Costs

When you rent an apartment, it is important to understand clearly what costs and responsibilities you will have in addition to your monthly rent. Renters often pay a **security deposit.** This money is held by the landlord for unexpected costs, such as damages caused by the renter. Usually a written lease explains what types of repairs you are expected to make and whether or not you must pay your own utilities.

EXAMPLE Sue and Jill Ardant rent an apartment for $350 per month. When they sign the lease, they must pay first and last month's rent, plus a security deposit of $200. How much will they spend to rent the apartment?

First month's rent		Last month's rent		Security deposit		Total
$550	+	$550	+	$200	=	$1300

Solve.

1. Sharon found an apartment for $725 per month. To rent it, she will need to pay first month's rent and a $300 security deposit. Utilities will be $90 extra. What will she spend to rent the apartment the first month?

Answer ________________

2. Joe rents an apartment for $400 per month. He estimates that he will pay $50 per month for electricity and $75 per month for parking. How much will he be spending per month on rent, electricity, and parking?

Answer ________________

3. Bill and Jose decided to share an apartment that costs $700 a month. They also pay $80 a month for electricity, $30 for gas, $5 for water, and $7 for garbage pickup. If they share the cost equally, how much will each person pay a month?

Answer ________________

4. When Phyllis moved out of her apartment, she checked her lease and found that she needed to have the carpets cleaned in order to get back her security deposit of $400. The landlord charged her $45 for the cleaning. How much was left of the $400 after Phyllis paid for the carpet cleaning?

Answer ________________

5. The average cost of heating a one-bedroom apartment in Chicago is $80 per month during the 6-month winter season. Julie wants to rent an apartment for a year. Which is less expensive, one with heat included for $450 per month, or one without heat included for $390 per month?

Answer ________________

6. When Heather rented her new apartment, she agreed to have the living room repainted every 5 years. The painting costs $300. How much extra per month is Heather spending for the apartment if she stays for 5 years?

Answer ________________

PROBLEM-SOLVING STRATEGY

Use a Formula

A formula shows a relationship between quantities. For example, the formula $I = p \times r \times t$ shows that interest is the product of the principal, rate, and time. Formulas can be used to solve problems.

A person who owns a home must pay real estate taxes and fire insurance for the home. Some mortgage companies require a portion of taxes and insurance to be paid along with the regular mortgage payment. The total monthly payment is often referred to by the initials PITI, for Principal, Interest, Taxes, and Insurance.

If a family is thinking of buying a home, they should talk with a mortgage lender to determine how large of a monthly mortgage payment they can afford. The maximum payment is based on gross income and debts. Some banks use a formula such as this.

MORTGAGE PAYMENT AFFORDABILITY FORMULA

Choose A or B, whichever is less:

(A) PITI = 26% × gross income

(B) PITI = (32% × gross income) − other debt payments

PITI = principal + interest + real estate taxes + fire insurance

Read the problem.

Rose is applying for a mortgage. Her gross income is $2800 per month and she has debt payments of $240 per month. How much can she afford for PITI each month?

Use the formula.

(A) PITI = 26% × $2800 = $728

(B) PITI = (32% × $2800) − $240
= $896 − $240
= $656

Solve the problem.

Choose the amount which is less, $656. So, Rose can afford $656 per month for principal, interest, taxes, and insurance.

Solve. Use the formulas shown above.

1. Colleen earns $2200 per month. She does not have any debts. How much can she afford to spend each month for PITI?

Answer ____________________

2. Travis earns $2400 per month. He has a car payment of $190 per month and no other debts. How much can he afford for housing payments (PITI)?

Answer ____________________

Solve.

3. Olivia earns $3700 per month. She has a car payment of $250 and other debt payments of $120. How much can she afford to spend each month for PITI? Use the formula on page 126.

Answer ________________

4. Mr. Ogawa earns $2600 per month. Mrs. Ogawa works part-time and earns $1500 per month. They have monthly debt payments of $230. How much can they afford to spend each month for PITI?

Answer ________________

5. Most people can afford a house that costs about twice their yearly income. About what income is needed to buy a house for $88,000?

Answer ________________

6. If a family has gross income of $52,000, about how much can they afford to pay for a house?

Answer ________________

7. The interest formula $I = p \times r \times t$, may also be expressed as $p = \frac{I}{r \times t}$. Find the principal if the rate is 9% and the interest for one year is $5040.

Answer ________________

8. The interest formula, $I = p \times r \times t$, may also be expressed as $t = \frac{I}{p \times r}$. Find the time if the rate is 10%, the principal is $35,000, and the interest is $1750.

Answer ________________

9. Frederick Reese is getting a mortgage. The payment for principal and interest each month will be $755. The real estate taxes are $1070 per year. The fire insurance is $250 per year. One-twelfth of the taxes and insurance will be added to Frederick's payment each month. How much will Frederick pay for PITI each month?

Answer ________________

10. Amanda Dixon is getting a mortgage. The payment for principal and interest each month will be $808. The real estate taxes are $1288 per year. The fire insurance is $260 per year. One-twelfth of the taxes and insurance will be added to Amanda's payment each month. How much will she pay for PITI each month?

Answer ________________

Using Area and Perimeter Formulas

EXAMPLE 1 What is the area of the front face of this brick?

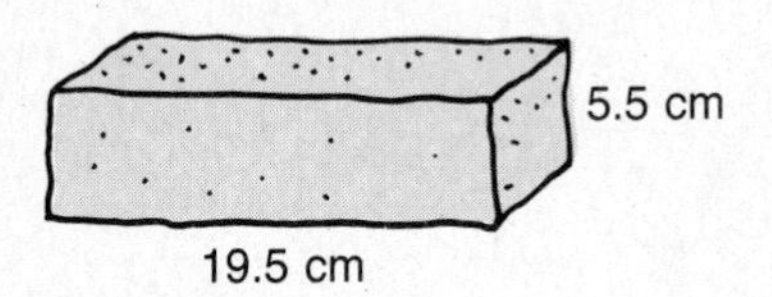

The area of a rectangle is the length times the width.

$A = l \times w$

$A = 19.5 \times 5.5$

$A = 107.25$ square centimeters (cm^2)

What is the perimeter of the front face of the brick?

The perimeter is the sum of the lengths of the sides.

$P = l + w + l + w$ or $2l + 2w$

$P = 2(19.5) + 2(5.5)$

$P = 50$ centimeters

EXAMPLE 2 What is the area of one triangular face of this roof?

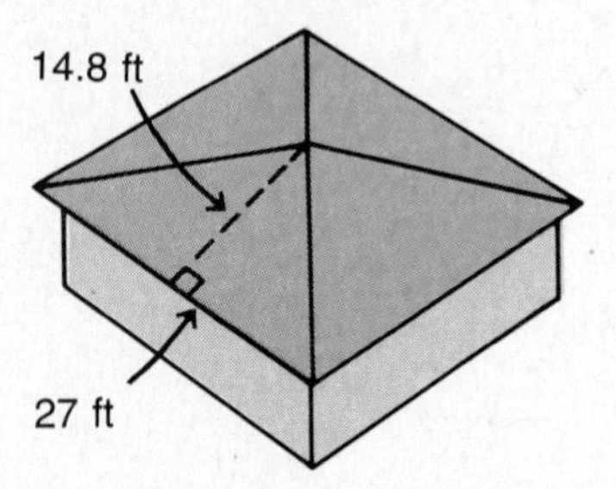

The area of a triangle is $\frac{1}{2}$ times the base times the height.

$A = \frac{1}{2} \times b \times h$

$A = \frac{1}{2} \times 27 \times 14.8$

$A = 199.8$ square feet (sq ft or ft^2)

What is the perimeter of the triangle above? The other two sides are 20 feet each.

$P = s + s + s$

$P = 20 + 20 + 27$

$P = 67$ feet

Solve.

1. Andrew Willis has a garden 8 meters long and 6 meters wide. If a gate is 0.5 meter wide, how many meters of fence will he need to enclose the garden?

Answer ______________

2. Mary Mathews wishes to put a strip of trim, called molding, around the bottom and the top of her living room walls. The room is 16 feet by 14 feet. How many feet of molding are needed?

Answer ______________

3. Kevin Sullivan wished to have his yard sodded. The yard is 60 ft by 40 ft. Estimate how many square feet of sod he needs.

Answer ______________

4. Andrea bought some broadloom carpeting. She bought a piece measuring 9 feet by 15 feet. How many square feet did she buy?

Answer ______________

Solve.

5. Find the cost of laying a concrete walk 50 feet long and 4 feet wide. The concrete costs $1.75 per square foot.

Answer ____________

6. A highway cut off a corner of Jean Marino's property. The piece cut off formed a triangle whose base was 25 meters and whose height was 30 meters. How many square meters were in the triangle?

Answer ____________

7. Nina Hatfield wants to build a fence all the way around a field. The field is 100 meters long and 80 meters wide. How many meters of fence will she need if she allows two meters for the gate?

Answer ____________

8. A baseboard is a strip of trim along the bottom of a wall. How many feet of baseboard will be needed for a playroom measuring 28 feet by 22 feet if you allow 6 feet for the door openings?

Answer ____________

9. Brenda Hill wants to buy carpeting for a room which measures 18 feet by 14 feet. Estimate how many square yards of carpet she will need.

Answer ____________

10. How much will Brenda Hill have to pay for the carpet at $10 per square yard. (Refer to your answer to Exercise 9.)

Answer ____________

11. A dining room floor is 40 feet wide and 90 feet long. Estimate how much it will cost to have the floor sanded at $1.25 a square yard. (There are 9 square feet in a square yard.)

Answer ____________

12. A rug that measures 9 ft by 12 ft sells for $144. How much does it cost per square yard?

Answer ____________

13. Ben Hendrickson is building a triangular patio. The sides measure 12 feet, 16 feet, and 20 feet. What is the perimeter of the patio?

Answer ____________

14. During a 10%-off sale, Becky Schmidt bought enough carpet to cover an area 18 ft by 12 ft. The carpet regularly sold for $20 per square yard. Estimate how much Becky paid for the carpet.

Answer ____________

Volume and Surface Area

The volume of the concrete block is the number of cubic units that take up the same amount of space. The concrete block is 26 cm long, 10 cm wide, and 11 cm high.

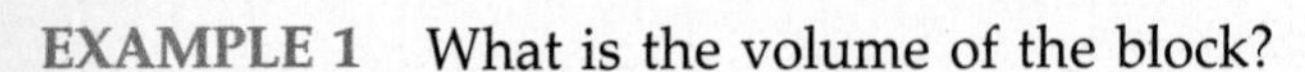

EXAMPLE 1 What is the volume of the block?

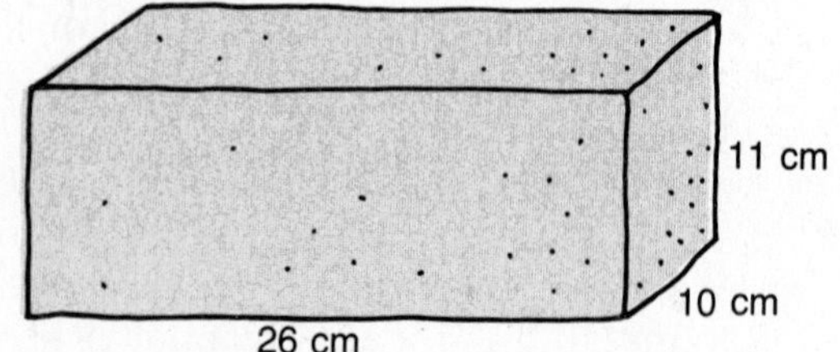

Length		Width		Height		Volume
26	$\times$	10	$\times$	11	=	2860

The volume is 2860 cubic centimeters (cm^3).

EXAMPLE 2 What is the surface area of the block?

Surface area = sum of areas of surfaces

Front and back:	$26 \times 11 = 286$ cm^2 each
Sides:	$10 \times 11 = 110$ cm^2 each
Top and bottom:	$10 \times 26 = 260$ cm^2 each

Add. $286 + 286 + 110 + 110 + 260 + 260 = 1312$

The surface area of the block is 1312 cm^2.

Solve.

1. For the basement of a house, a hole 30 ft wide, 40 ft long, and 10 ft deep was made. How many cubic feet of dirt were removed?

Answer ____________

2. A shelf will be covered with veneer on all 6 sides. The shelf is 48 inches long, 12 inches wide, and 1 inch thick. What is the surface area of the shelf?

Answer ____________

3. Terese Harper laid fresh topsoil 4 inches deep on her garden. The garden is 60 ft by 54 ft. How many cubic feet of soil did she need? (First change 4 inches to a fraction of a foot.)

Answer ____________

4. For a basement, dirt must be removed from a space that is 24 ft by 21 ft by 12 ft. Allowing 4 cubic yards to a load, how many loads of dirt must be removed? (First change the three dimensions to yards.)

Answer ____________

5. A garden shop sells firewood by the cord, which is 128 cubic feet. How many cords of wood are there in a stack which measures 12 ft by 18 ft by 6 ft?

Answer ____________

6. George Baker has a house that is about 30 ft wide, 100 ft in length, and 40 ft in height to the beginning of the roof. To estimate the amount of paint he needs, find the area in square feet of the four sides of the house.

Answer ____________

Calculator Applications

Measurements are not exact. For example, suppose the length of a rectangle is measured as 17.4 cm and the width is measured as 3.8 cm. These measurements are accurate to tenths of centimeters, but the actual length and width may be a little shorter or longer. The actual length could be between 17.35 cm and 17.45 cm. The actual width may be between 3.75 and 3.85 cm.

If you multiply 17.4 by 3.8 to find the area of the rectangle, not all of the digits in the product will be accurate. Compare the results at the right. Notice that the actual area may vary from 65.06 to 67.18. Because the digits of 66.12 are not all accurate, it makes more sense to report the product as 66 cm^2 instead of 66.12 cm^2.

Length		Width		Area
17.35	×	3.75	=	65.0625
17.4	×	3.8	=	66.12
17.45	×	3.85	=	67.1825

Here is an easy way to decide how to round the product when finding area. Look at the factors to see which one has fewer **significant digits**. Significant digits are the nonzero digits, plus zeros that are between nonzero digits. A measurement is less *accurate* if it has fewer significant digits. When you multiply measurements, the product should be rounded to the same number of significant digits as the less accurate factor.

EXAMPLE Find the area of a rectangle that measures 60.2 m by 137.5 m.

60.2 × 137.5 = 8277.5

3 significant digits 4 significant digits

If you round 8277.5 to 3 significant digits, the result is 8280. So the area of the rectangle should be reported as 8280 m^2.

Write the number of significant digits in each number.

1. 25 ______ **2.** 47.1 ______ **3.** 10.7 ______ **4.** 40 ______

5. 0.7 ______ **6.** 9.408 ______ **7.** 2300 ______ **8.** 0.09 ______

The dimensions of a rectangle are given. Use a calculator to find the area. Round the product so it has the same number of significant digits as the less accurate factor.

9. 7.5 m × 8.2 m ______ **10.** 0.8 m × 0.9 m ______

11. 16.5 ft × 203 ft ______ **12.** 47 in. × 30.5 in. ______

13. 62 cm × 0.6 cm ______ **14.** 132.8 mi × 50.7 mi ______

Insulation and Roofing

Houses are built with **insulation** in the walls to prevent heat loss. Insulation for some buildings may be soft, flexible blankets made of fiberglass. The insulation comes in several widths so that it fits exactly between the boards in a wall or ceiling.

EXAMPLE 1 Estimate the cost of insulating the ceiling of a house that is 30 ft wide and 50 ft long. One package of insulating material covers 48 square feet and costs $30. Estimate the number of packages needed by using the formula for area of a rectangle, $A = l \times w$.

Area of ceiling:	$30 \times 50 = 1500$ sq ft
Number of packages needed:	$1500 \div 48 = 31.25$ or about 32
Cost:	32 packages × $30 = $960

Solve.

1. Susan Winters is insulating a ceiling that is 40 ft wide and 60 ft long. The insulation costs $55 per package and each package covers 64 sq ft. Estimate how much the insulation will cost.

Answer ____________________

2. Fred Drew decided to insulate one wall of his den. He chose insulation that costs $40 a package. Each package covers 48 sq ft. The wall is 3 yards by 5 yards. Estimate the insulation cost.

Answer ____________________

Roofing shingles are sold by the square. A **square** of shingles covers a 100-square-foot section of the roof.

EXAMPLE 2 Estimate the cost of shingling a roof which measures 2400 square feet. The shingles cost about $40 per square and the labor charge for installation is $50 per square.

Number of squares needed:	$2400 \div 100 = 24$
Cost of shingles:	24 squares × $40 = $960
Cost of labor:	24 squares × $50 = $1200
Estimated total cost:	$960 + $1200 = $2160

Solve.

3. Estimate the cost of roofing a building with 3000 square feet of roof. Shingles cost $35 per square. The labor charge is $30 per square.

Answer ____________________

4. Estimate the cost of roofing the building shown at the right. Shingles cost $50 per square. Labor is $30 per square.

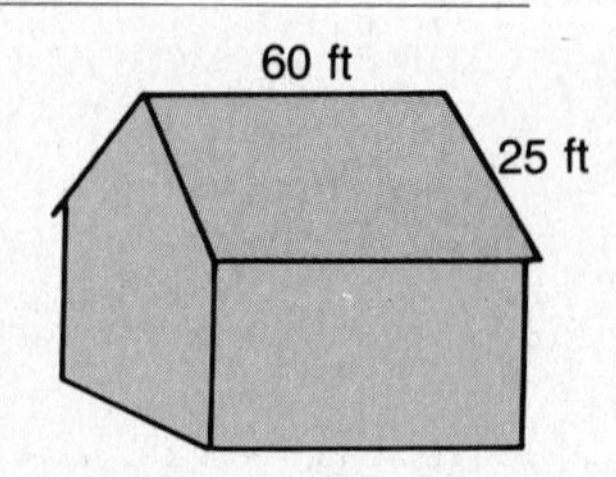

Answer ____________________

Floors and Plumbing

Tiles are used for floors in kitchens, bathrooms, and other rooms of a house. The tiles come in different materials and are sold either by the carton or individually.

EXAMPLE Estimate the cost of putting a new tile floor in the kitchen that measures 15 ft by 12 ft. One carton of tiles costs $20 and will cover 50 sq ft. However, you cannot buy part of a carton.

Area of floor: $12 \times 15 = 180$ square feet
Number of cartons: $180 \div 50 = 3.6$

You will need 4 cartons at $20 each. The cost will be about $80.

Solve.

1. Tom plans to tile a bathroom that measures 6 ft by 8 ft. One carton of tiles costs $35 and will cover 40 square feet. How much will the tiles cost?

Answer ______________________

2. The floor plan of a kitchen is shown below. Find the cost of tiling the floor. The tiles cost $30 per carton. Each carton covers 30 square feet.

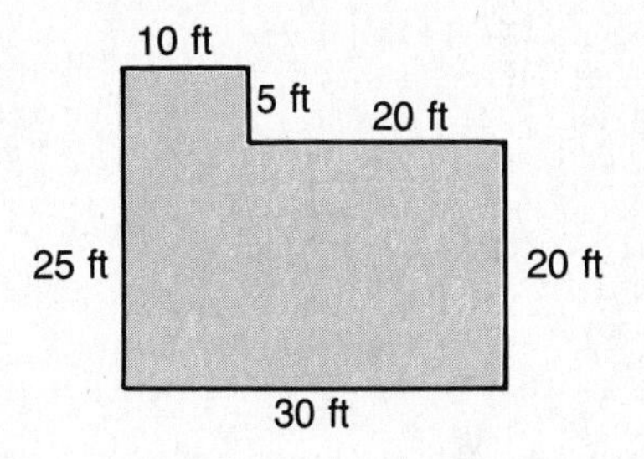

Answer ______________________

3. Susan rents an apartment and wants to put a new tile floor in the bathroom. Her landlady will pay for the tiles if Susan does the work. The tiles are 6 in. by 6 in. and the floor measures 10 ft by 8 ft. How many tiles will Susan need? (First find the number of tiles along the length and width.)

Answer ______________________

Plumbing repairs can be expensive, and if not done properly other damages can occur. It is important to hire a plumber or plumbing company that will guarantee the work.

4. Gina Martin hired a plumber to fix four clogged drains. The basic service charge was $42 for the first hour. The cost for additional time was $9.50 per 15 minutes. If the job took $1\frac{1}{2}$ hours, what was the cost?

Answer ______________________

5. Joy and Mark Jacobs hired a plumber to replace a drain and install a new garbage disposal. The plumber worked $1\frac{3}{4}$ hours, and charged $98.20 for parts. What was the total cost? (Use the rates from Exercise 4.)

Answer ______________________

PROBLEM-SOLVING STRATEGY

Make a Drawing

When you read a problem, you might not know at once how to solve it. When that happens, a drawing may help you decide what to do. A rough sketch is usually good enough. Make sure to label the drawing with the facts given in the problem.

Read the problem.

Tom's bedroom is 10 feet 4 inches by 14 ft 6 inches. He gets a rug that measures 9 feet by 12 feet to put in the room. If he places the rug in the center, what is the distance to the wall on each end of the rug? What is the distance to the wall on each side?

Make a drawing.

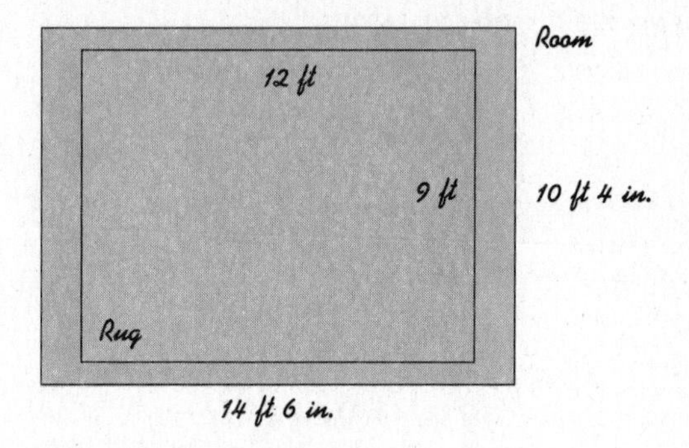

Solve the problem.

Since the rug is in the center of the room, half of the uncovered distance is at each end of the room. Subtract to find the length and width not covered. Then divide each result by 2 to find each distance from the rug to the wall.

Distance to the wall:

End: (14 ft 6 in. − 12 ft) ÷ 2
= 2 ft 6 in. ÷ 2
= 1 ft 3 in.

Side: (10 ft 4 in. − 9 ft) ÷ 2
= 1 ft 4 in. ÷ 2
= 16 in. ÷ 2
= 8 in.

The distance from the rug to each end of the room is 1 ft 3 in. The distance from the rug to each side of the room is 8 in.

Solve.

1. One surface of a roof is shaped like a rectangle, 20 feet wide and 16 feet high. Draw a picture of the roof surface in the space at the right.

2. The shingles and roof boards need to be replaced. How many 4-ft by 8-ft pieces of plywood would cover the roof?

Answer ________________

Solve.

3. A room is 12 feet 2 inches wide and 15 feet 6 inches long. One of the longer sides of the room has two doorways each 32 inches wide. Make a drawing of the room in the space at the right.

4. How much molding is needed for the base of the four walls, not including the doorways?

Answer ______________________

5. The Dean's house was built on a rectangular lot that is 70 feet wide along the street. The lot is 120 feet deep. Make a drawing of the lot in the space at the right.

6. If Mr. Dean stands 12 feet from the street, how far is he from the back of the lot?

Answer ______________________

7. The base of the Dean's house is also shaped like a rectangle. The front of the house is 30 feet from the street and the depth of the house is 26 feet. Show this on your drawing above. What is the distance from the back of the house to the back of the lot?

Answer ______________________

8. The Dean's house is 44 feet wide, centered on the width of the lot. How far is it to the property line on each side?

Answer ______________________

9. Follow these directions to draw the patio on the Ellis's house. The drawing is started for you in the space at the right. Label the length of each side as 10 units. Then draw a line from point D in the opposite direction from C. Make the line 4 units long and label the endpoint E. Draw a straight line from E to A. The patio is the shape of polygon ABCDE.

A B

D C

10. What is the area of the triangle ADE?

Answer ______________________

11. What is the area of square ABCD? If each unit is a foot, what is the total area of the patio?

Answer ______________________

Building a Home

There are many costs associated with building a home. Some people hire an **architect** to design the home and draw construction plans. Also, most people hire a **contractor** to construct the house. Other costs are buying the lot, paying for utility service to the lot, and landscaping.

EXAMPLE Mr. Yazzie paid $1.70 per square foot for a lot that measured 60 feet by 130 feet. How much did he pay for the lot?

Total number of square feet: $60 \times 130 = 7800$ sq ft
Total cost: $7800 \times \$1.70 = \$13,260$.

Mr. Yazzie paid $13,260 for the lot.

Solve.

1. Mr. Yazzie decided to build a five-room, one-bath house on the lot. There will be two bedrooms, each 14 ft by 12 ft, a living room 18 ft by 20 ft, a dining room 12 ft by 15 ft, a kitchen 9 ft by 12 ft, and a bathroom 7 ft by 10 ft. Complete the chart at the right to find the total number of square feet in the house.

Room	Dimensions	Area
Bedroom	14 ft × 12 ft	
Bedroom		
Living room		
Dining room		
Kitchen		
Bathroom		
TOTAL		

2. The contractor said she could build the house for $70 per square foot of floor surface. What would the construction cost be? (Use total area from the chart.)

Answer ________________

3. The architect who drew the plans and supervised the construction charged a fee of 5% of the construction cost. How much was the fee?

Answer ________________

4. Mr. Yazzie wants to carpet the living room and dining room. At $17.50 per square yard, estimate how much it will cost.
(Hint: 1 square yard is 9 square feet.)

Answer ________________

5. The house plans show a basement that will be 24 ft wide, 30 ft long, and 9 ft deep. At $10 per cubic yard, what will be the cost to dig space for the basement?

Answer ________________

6. At 80 cents per square foot tile, what will be the cost of tile for the kitchen floor?

Answer ________________

7. Linoleum for the kitchen floor would cost $5.30 per square yard. How much cheaper would this be than the tile?

Answer ________________

Buying for the Home

When you move to a new home, you might need to buy new furniture and appliances. If you are a careful shopper, you can often find bargains that will save you money.

Solve.

1. When she moved into her new home, Janet bought new furniture which would have cost $1350. By paying cash, she was given a 5% discount. How much did she save?

Answer ____________

2. Edward found a store that was going out of business. He bought a television set originally priced at $425 at 25% off. How much did he save?

Answer ____________

3. Alex Cortex bought a new refrigerator. The usual price of it was $895. By paying cash he saved 10%. How much was saved?

Answer ____________

4. Tony Jefferson bought a refrigerator during a 20%-off sale. The refrigerator regularly sold for $575. How much did he pay?

Answer ____________

5. By paying cash Paula Ling was able to buy a $450 stove for $400.50. What percent did she save?

Answer ____________

6. Shana wants to buy a new air conditioner which sells for $550. If she waits until winter, she can buy it at a 20% discount. How much would it cost in the winter?

Answer ____________

7. Jerry bought a stove for $600. He paid $200 down and the balance, including interest, in 12 monthly installments of $40. How much did the stove cost, including interest?

Answer ____________

8. Carla Robins bought a $450 freezer and paid $50 down. The balance was to be paid in 6 monthly payments of $70.50 each. How much interest did she pay in all?

Answer ____________

Comparing Costs

When you purchase building supplies or appliances, price is not the only aspect to consider. Some items may have a longer warranty than others, or require less upkeep. The energy usage of appliances should also be considered.

EXAMPLE Marla is choosing paint for her living room. Brand A has a 10-year warranty and costs $12.99 per gallon. Brand B has a 5-year warranty and costs $8.99 per gallon. She decides to buy 2 gallons of paint. If 7% sales tax is added, what is the total cost for each brand? Which paint is the better buy?

	Brand A	**Brand B**
Cost for 2 gallons:	2 × $12.99 = $25.98	2 × $8.99 = $17.98
Sales tax:	$25.98 × 0.07 = $ 1.82	$17.98 × 0.07 = $ 1.26
Total:	$27.80	$19.24

Brand B costs less but might not last as long.

Solve.

1. Bonita spends about $1200 every five years to have her house painted. She got an estimate for $4500 to have vinyl siding put on the house. The siding has a 20-year warranty and will not need to be painted. She is planning to live in the house for at least 20 more years. How much money will she save by buying siding?

Answer ______________________

2. Richard Colson is comparing the cost of two brands of lawn fertilizer. Both are sold in 50-pound bags. Brand A costs $15 and covers 1250 sq ft. Brand B costs $13.50 and covers 675 sq ft. Which is the better buy? How much less does it cost per square foot?

Answer ______________________

3. Julie Kelso compares two refrigerators. Brand A costs $725 and has an estimated electrical use cost of $120 a year. Brand B costs $850 and has an estimated electrical use cost of $75 a year. If she plans to use the appliance for several years, which one is the better buy?

Answer ______________________

4. Peter Mangini compares two window air conditioners. Brand A has an Electrical Efficiency Rating (EER) of 8 and costs $365. Brand B has an EER of 12 and costs $550. Estimated yearly operating costs are $240 for Brand A and $160 for Brand B. If Peter plans to use the air conditioner for several years, which one is the better buy?

Answer ______________________

Unit 6 Review

Solve.

1. Susan Rubin bought a house for $130,000. She paid 25% down and took out a mortgage loan for the remaining money. How much was her loan?

Answer ______________________

2. Dick Hartman buys a condominium for $90,000 and makes a down payment of $30,000. His mortgage loan is for 15 years at 8%. The mortgage payment is $9.56 per thousand of the loan amount. What is his monthly payment?

Answer ______________________

3. Last month Ralph's electric meter read 2365 kilowatt-hours. This month it reads 3214 kilowatt-hours. At 7¢ per kilowatt-hour, what will Ralph's electric bill be for the month?

Answer ______________________

4. June and Bill Fredricks want to buy $80,000 worth of insurance for their new house. If the rate is $7.00 per $1000 of value, what will their insurance premium be?

Answer ______________________

5. Melissa Walker found an apartment for $530 per month. To move in, she will need to pay first and last month's rent, plus a security deposit of $350. What will she spend to move into the apartment?

Answer ______________________

6. Mike Rojas put in a sidewalk 5 feet by 40 feet. If it cost him $2.50 per square foot, how much did he pay for the sidewalk?

Answer ______________________

7. Workers dug a hole for the basement of Sharon Anthony's house. The hole was 40 feet wide, 50 feet long, and 9 feet deep. How many cubic feet of dirt were removed from the hole?

Answer ______________________

8. Estimate the cost of roofing a building if the roof area is 3600 square feet. The shingles will cost $40 per square and the labor charge is $35 per square. (One square of shingles covers 100 square feet.)

Answer ______________________

9. Lynnette needed to hire a plumber to install a new sink. The plumber charged for 2 hours and 45 minutes of labor. The rate was $40 for the first hour and $8 for each additional 15 minutes. How much did Lynnette pay for the labor charge?

Answer ______________________

10. By paying cash instead of charging, Carolyn Suni bought a $350 television for $315. What percent did she save?

Answer ______________________

Federal Income Withholding Tax

If you are employed, your employer is probably required to withhold money from each paycheck for federal income tax. The amount of tax withheld depends on your salary, the length of the pay period (weekly, monthly, etc.) whether you are single or married, and the number of withholding allowances you claim. Withholding allowances usually include yourself, your spouse (if married), and your dependent children.

When you begin a new job, the employer will ask you to fill out a Form W-4 like the one below. With that information, your employer will use a table like the partial table at the right to compute how much to withhold from your paycheck.

SINGLE Persons–**WEEKLY** Payroll Period

And the wages are–		And the number of withholding allowances claimed is–				
At least	But less than	0	1	2	3	4
		The amount of income tax to be withheld shall be–				
240	250	34	28	22	16	11
250	260	35	29	24	18	12
260	270	37	31	25	19	14
270	280	38	32	27	21	15
280	290	40	34	28	22	17
290	300	41	35	30	24	18
300	310	43	37	31	25	20
310	320	44	38	33	27	21
320	330	46	40	34	28	23
330	340	47	41	36	30	24
340	350	49	43	37	31	26
350	360	50	44	39	33	27
360	370	52	46	40	34	29
370	380	53	47	42	36	30
380	390	56	49	43	37	32
390	400	58	50	45	39	33
400	410	61	52	46	40	35
410	420	64	53	48	42	36
420	430	67	56	49	43	38
430	440	70	59	51	45	39
440	450	72	62	52	46	41
450	460	75	64	54	48	42
460	470	78	67	56	49	44
470	480	81	70	59	51	45
480	490	84	73	62	52	47

Cut here and give the certificate to your employer. Keep the top portion for your records.

Form **W-4** Department of the Treasury Internal Revenue Service

Employee's Withholding Allowance Certificate

► For Privacy Act and Paperwork Reduction Act Notice, see reverse.

OMB No. 1545-0010 19—

1 Type or print your first name and middle initial — Last name: RUSSELL BRADY

2 Your social security number: 328 - 01 - 4346

Home address (number and street or rural route): 1245 HOME AVE

City or town, state, and ZIP code: BERWYN, IL 60402

3 Marital Status: ☑ Single ☐ Married ☐ Married, but withhold at higher Single rate.
Note: *If married, but legally separated, or spouse is a nonresident alien, check the Single box.*

4 Total number of allowances you are claiming (from line G above or from the Worksheets on back if they apply) . . . 4 1

5 Additional amount, if any, you want deducted from each pay . . . . 5 $

6 I claim exemption from withholding and I certify that I meet **ALL** of the following conditions for exemption:

- Last year I had a right to a refund of **ALL** Federal income tax withheld because I had **NO** tax liability; **AND**
- This year I expect a refund of **ALL** Federal income tax withheld because I expect to have **NO** tax liability; **AND**
- This year if my income exceeds $500 and includes nonwage income, another person cannot claim me as a dependent.

If you meet all of the above conditions, enter the year effective and "EXEMPT" here . . . . . ► 6 19

7 Are you a full-time student? (**Note:** *Full-time students are not automatically exempt.*) . . . . 7 ☐Yes ☐No

Under penalties of perjury, I certify that I am entitled to the number of withholding allowances claimed on this certificate or entitled to claim exempt status.

Employee's signature ► Russell Brady **Date ►** 5-31 , 19—

8 Employer's name and address (**Employer:** Complete 8 and 10 **only if sending to IRS**) | 9 Office code (optional) | 10 Employer identification number

EXAMPLE Russell Brady filled out the form to show that he is single and is claiming 1 withholding allowance (box 3). (If Russell had wanted to have an additional amount withheld each week, he could have written the amount on Line 5.) Russell Brady's salary is $274 per week. How much money is withheld from Russell's salary each week for federal income taxes?

In the table, 274 is between 270 and 280. Read across the row to the column labeled 1 for one withholding allowance. The amount withheld is $32.

For each exercise, the number of allowances and weekly salary of a single person are listed. Use the partial table on page 140 to find the amount withheld from each paycheck.

1. 2 allowances, $314 ____________

2. 0 allowances, $405 ____________

3. 1 allowance, $289 ____________

4. 3 allowances, $347 ____________

For each exercise, the number of allowances and weekly salary of a married person are listed. Use the partial table below to find the amount withheld from each paycheck.

5. 3 allowances, $325 ____________

6. 2 allowances, $483 ____________

7. 4 allowances, $505 ____________

8. 1 allowance, $232 ____________

9. 0 allowances, $401 ____________

10. 2 allowances, $308 ____________

11. Jan is married and claims 2 allowances. She earns $181 per week, and $7 is withheld from each paycheck. Suppose Jan gets a new job and earns exactly twice as much as before. How much would be withheld? Is the withholding amount twice as much as before?

Answer ____________________

12. Suppose Jan would like to have $560 more deducted during one year. About how much should be deducted from each of her 52 paychecks?

Answer ____________________

13. Rita and Jack are married. Rita earns $423 per week and claims 3 allowances. Jack works parttime, earns $156 per week, and claims 0 allowances. What is the total withheld for the couple each week? (Find each withholding amount separately.)

Answer ____________________

14. Refer to Exercise 13. Suppose Rita and Jack provide new W-4 Forms. Rita claims 0 allowances and Jack claims 3. How much more or less federal tax will be withheld for the couple each week?

Answer ____________________

MARRIED Persons—**WEEKLY** Payroll Period

And the wages are—		And the number of withholding allowances claimed is—				
At least	But less than	0	1	2	3	4
		The amount of income tax to be withheld shall be—				
135	140	11	6	0	0	0
140	145	12	6	1	0	0
145	150	13	7	1	0	0
150	155	14	8	2	0	0
155	160	14	9	3	0	0
160	165	15	9	4	0	0
165	170	16	10	4	0	0
170	175	17	11	5	0	0
175	180	17	12	6	0	0
180	185	18	12	7	1	0
185	190	19	13	7	2	0
190	195	20	14	8	2	0
195	200	20	15	9	3	0
200	210	22	16	10	4	0
210	220	23	17	11	6	0
220	230	25	19	13	7	1
230	240	26	20	14	9	3
240	250	28	22	16	10	4
250	260	29	23	17	12	6
260	270	31	25	19	13	7
270	280	32	26	20	15	9
280	290	34	26	22	16	10
290	300	35	29	23	18	12
300	310	37	31	25	19	13
310	320	38	32	26	21	15
320	330	40	34	28	22	16
330	340	41	35	29	24	18
340	350	43	37	31	25	19
350	360	44	38	32	27	21
360	370	46	40	34	28	22
370	380	47	41	35	30	24
380	390	49	43	37	31	25
390	400	50	44	38	33	27
400	410	52	46	40	34	28
410	420	53	47	41	36	30
420	430	55	49	43	37	31
430	440	56	50	44	39	33
440	450	58	52	46	40	34
450	460	59	53	47	42	36
460	470	61	55	49	43	37
470	480	62	56	50	45	39
480	490	64	58	52	46	40
490	500	65	59	53	48	42
500	510	67	61	55	49	43
510	520	68	62	56	51	45
520	530	70	64	58	52	46
530	540	71	65	59	54	48
540	550	73	67	61	55	49
550	560	74	68	62	57	51
560	570	76	70	64	58	52
570	580	77	71	65	60	54
580	590	79	73	67	61	55
590	600	80	74	68	63	57
600	610	82	76	70	64	58
610	620	83	77	71	66	60

Social Security Tax

The Federal Insurance Contributions Act (FICA) requires employers to withhold a portion of the pay of most employees for **social security** and **Medicare** taxes. In 1994, the rate was 7.65%. The employer must contribute an amount that equals the employee's contribution and send both amounts to the federal government. The money is used to pay retirement benefits, disability benefits, medical-cost benefits, and survivor's benefits to qualified persons.

EXAMPLE Natalie earns $429.45 per week. Using a rate of 7.65%, how much social security tax is withheld from each week's paycheck?

7.65% of $429.45 = 0.0765 × $429.45
= $32.852925
= $32.86 Round up to the next cent.

People who are self-employed must pay 15.3% of their profit for social security taxes. The rate is higher because a self-employed person's contribution is not being matched by an employer's contribution. A self-employed person is required to send these taxes directly to the government in quarterly payments (every three months), along with federal income tax payments.

Find the amount of social security tax withheld for the given gross weekly pay. Use a tax rate of 7.65%.

1. $514.56 ____________

2. $202.83 ____________

3. $347.58 ____________

4. $798.45 ____________

5. Sue's salary is $33,000 per year. How much is withheld each year for social security tax?

Answer ____________

6. Larry earns $42,750 per year. What is the total of Larry's social security contribution and his employer's matching contribution?

Answer ____________

7. Greg is self-employed and made a profit of $37,341 in one year. How much social security tax did he owe for the year? Use a rate of 15.3%.

Answer ____________

8. Louisa is self-employed and expects a profit of $12,500 in the first quarter of the year. How much social security tax should she pay for the quarter? Use a rate of 15.3%.

Answer ____________

Income Reports

In January of each year, your employer will send you a Wage and Tax Statement Form, called a **Form W-2.** The W-2 form tells you how much you earned the previous year and how much was withheld for federal, state, and local income tax. It also shows the amount withheld for social security (FICA).

Form 1099 is a similar form that is used to report miscellaneous income. If your bank account earned interest, your bank sends you a Form 1099 showing how much interest was earned. If you are self-employed and companies paid you for services, the companies will send 1099 forms showing the total paid to you during the year.

OMB No. 1545-0008

1 Control number 50040N

4 Employer's State number

FORM W-2 1993 Wage and Tax Statement

Copy D for employer

Department of the Treasury Internal Revenue Service

2 Employer's name, address, and ZIP code

Merchant's Inc.
2345 State Street
New York, NY 10017

3 Employer's identification number

5 Statutory employee | Deceased | Pension plan | Legal rep. | 942 emp. | Subtotal | Deferred compensation | Void

6 Allocated tips

7 Advance EIC payment

8 Employee's social security number 222-00-3333

9 Federal income tax withheld $4,872

10 Wages, tips, other compensation $32,150

11 Social security tax withheld

12 Employee's name

Marcia Murray
1234 Pleasant Ave.
Anytown, USA

Employee's address and ZIP code

13 Social security wages

14 Social security tips

16

16a Fringe benefits incl. in Box 10

17 State income tax $3,135

18 State wages, tips, etc.

19 Name of State

20 Local income tax

21 Local wages, tips, etc.

22 Name of locality

CENTER SEPARATE FORM. DELEAVE EACH SIDE, BURST AS REQUIRED. IF COPY A IS BURST, IT IS DONE AS AN 11" LENGTH.

EXAMPLE The W-2 form above shows Marcia Murray's wages for a year. She also received a 1099 form from the bank that showed interest earnings of $359. What was Marcia's gross income?

Salary		Interest income		Gross income
$32,150	+	$359	=	$32,509

Find the total of wages and interest income for each person.

	Name	Wages from W-2	Interest from 1099	Total
1.	P. Baty	$24,901	$389	
2.	C. Thomas	$12,073	$67	
3.	W. Myers	$34,289	$297	
4.	M. Anders	$50,388	$784	

5. Sam Snider's W-2 form showed that $834 was withheld for state tax and $490 was withheld for local tax. How much was withheld for state and local tax?

Answer ______________

6. A W-2 form shows that a person received wages of $24,952. If $998.08 was withheld for state taxes, what percent of wages were withheld for state taxes?

Answer ______________

PAYING TAXES

Form 1040EZ

By law, most people who earn money must prepare an income tax form by April 15 of each year and send it to the Internal Revenue Service, or IRS. Each tax form includes instructions for computing **taxable income.** Tables are included for figuring the **tax liability** based on taxable income and filing status. The **filing status** can be single, married filing separately, married filing jointly, or head of household.

Some of the tax liability is already paid by withholding tax, indicated on the W-2 form. If the tax liability is greater than the amount withheld, the taxpayer owes the IRS an amount called the **balance due.** If the tax liability is less than the amount withheld, the IRS will return the extra money as a **tax refund.** All amounts on tax forms can be rounded to the nearest dollar.

New tax forms are distributed each year by the IRS. There are several different forms available. The page at the right shows a sample of **Form 1040EZ,** the easiest form to use. This form is designed for single persons with no dependents, who earned less than $50,000 income and less than $400 in interest and dividends.

Linn Wong is single and has no dependents. She can use a 1040EZ Form as shown on the page at the right. Use information from the section of her W-2 form shown and the 1040EZ to complete the exercises below.

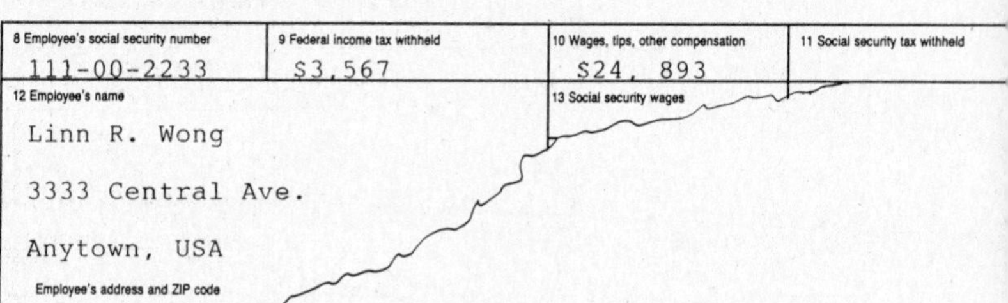
8 Employee's social security number: 111-00-2233
9 Federal income tax withheld: $3,567
10 Wages, tips, other compensation: $24, 893
11 Social security tax withheld
12 Employee's name
Linn R. Wong
3333 Central Ave.
Anytown, USA
Employee's address and ZIP code
13 Social security wages

1. Write Linn's name, address, and social security number in the spaces provided at the top of Form 1040EZ.

2. Linn's total wages are shown in Box 10 of her W-2 form. She also earned $148 interest on her savings account. Write these amounts on lines 1 and 2 of the 1040EZ. Then add and write the total on line 3. This is Linn's **adjusted gross income.**

3. Since Linn is not claimed as a dependent of someone else, check the box marked "No" and write 3,000 for her **standard deduction** on line 4. Subtract. Then write 2,000 for a **personal exemption** on line 6. Subtract. The remaining income is her **taxable income,** line 7.

4. Box 9 of Linn's W-2 form shows how much tax was withheld. Write this amount on line 8. Linn looked up her tax liability in the tax tables. It is $3287. Write this amount on line 9. Since line 8 is larger than line 9, subtract to find her refund.

5. Suppose Linn's employer had withheld $2,960 in federal tax. Would Linn get a refund (line 10) or owe (line 11)? How much?

   Answer ____________________

To complete the tax return, Linn would sign her name, write the date, and attach a copy of her W-2 form. If she owed money she would also attach a check or money order. Then she would make a copy of the completed form 1040EZ for her records and send the original to the address provided.

You may wish to make copies of this tax form to use for exercises on other pages. This form is a sample only. Because tax rules change frequently, new tax forms are distributed each year by the IRS.

Department of the Treasury - Internal Revenue Service

Form **1040EZ** **Income Tax Return for Single filers with no dependents** **19—** OMB No. 1545-0675

Name & address

Use the IRS mailing label. If you don't have one, please print.

LABEL HERE

Print your name above (first, initial, last)

Present home address (number, street, and apt. no.). (If you have a P.O. box, see back.)

City, town, or post office, state, and ZIP code

Please print your numbers like this: 0 1 2 3 4 5 6 7 8 9

Your social security number

Please read the instructions on the back of this form. Also, see page 13 of the booklet for a helpful checklist.

Presidential Election Campaign Fund
Do you want $1 to go to this fund? **Note:** *Checking "Yes" will not change your tax or reduce your refund.* ▶ Yes No

Dollars Cents

Report your income

Attach Copy B of Form(s) W-2 here

1 Total wages, salaries, and tips. This should be shown in Box 10 of your W-2 form(s). (Attach your W-2 form(s).) 1

2 Taxable interest income of $400 or less. If the total is more than $400, you cannot use Form 1040EZ. 2

3 Add line 1 and line 2. This is your **adjusted gross income.** 3

4 Can your parents or someone else claim you on their return?
☐ **Yes.** Do worksheet on back; enter amount from line E here.
☐ **No.** Enter 3,000 as your standard deduction. 4

Note: You must check Yes or No.

5 Subtract line 4 from line 3. If line 4 is larger than line 3, enter 0. 5

6 If you checked the "Yes" box on line 4, enter **0**.
If you checked the "No" box on line 4, enter **1,950.**
This is your **personal exemption.** 6

7 Subtract line 6 from line 5. If line 6 is larger than line 5, enter 0. This is your **taxable income.** 7

Figure your tax

8 Enter your Federal income tax withheld from Box 9 of your W-2 form(s). 8

9 Use the **single** column in the tax table on pages 37–42 of the Form 1040A/1040EZ booklet to find the **tax** on the amount shown on **line 7** above. Enter the amount of tax. 9

Refund or amount you owe

Attach tax payment here

10 If line 8 is larger than line 9, subtract line 9 from line 8. Enter the **amount of your refund.** 10

11 If line 9 is larger than line 8, subtract line 8 from line 9. Enter the **amount you owe.** Attach check or money order for the full amount, payable to "Internal Revenue Service." 11

Sign your return

I have read this return. Under penalties of perjury, I declare that to the best of my knowledge and belief, the return is true, correct, and complete.

Your signature Date

For IRS Use Only—Please do not write in boxes below.

Practice with Form 1040EZ

Suppose you are single and your parents or someone else *can* claim you on their return. If you had interest income, you must file a return if your total income was $500 or more. If you did *not* have interest income, you must file a return if your total wages were $3000 or more. You *cannot* take a personal exemption. (See line 6 of 1040EZ.)

On line 4 of Form 1040EZ, you can deduct from $500 to $3000 of your income. A worksheet for finding the exact deduction is provided on the back of Form 1040EZ. If your taxable income is less than $18,000, you can estimate the amount of tax by multiplying the taxable income by 15%. The exact amount of tax would be found by using a tax table.

Complete each column of the chart. Line 10 or 11 in each column should be blank. Refer to a copy of Form 1040EZ.

	1.	2.	3.	4.	5.
NAME OF TAXPAYER	**J. Sullivan**	**A. Rosner**	**B. Vincent**	**M. Riga**	**C. Sawada**
Line 1 Wages	$1307	$4589	$21,760	$16,304	$465
Line 2 Interest	43	0	352	185	290
Line 3 Adj. gross income	*1350*				
Line 4 Deduction	1307	3000	3000	3000	500
Line 5 Line 3 – line 4	*43*				
Line 6 Personal exemption	0	0	2000	2000	0
Line 7 Taxable income	*43*				
Line 8 Tax withheld	0	264	2691	1582	0
Line 9 Tax (Use 15% of line 7.) *Round to dollars.*	*6*				
Line 10 Refund					
Line 11 Amount you owe	*$6*				

6. Lori Raymas is single and is not claimed on anyone else's tax return. She earned wages of $18,394 and earned no interest. Her employer withheld $1980 for federal tax.

Her taxable income is ____________

Her tax liability (15%) is ____________

The amount she owes is ____________

7. Luke is single and is not claimed on anyone else's tax return. He earned $13,487 from one job and $5324 from another. The federal taxes withheld were $1279 and $572.

His taxable income is ____________

His tax liability (15%) is ____________

The amount he owes is ____________

PAYING TAXES

Calculator Applications

The memory keys of a calculator are helpful when a computation requires several steps. Compare these memory keys with the ones on your calculator. Are the keys labeled the same way?

M+ Adds the number in the display to the number in the memory.

M− Subtracts the number in the display from the number in the memory.

MR Recalls and displays the number that is in memory. (MR stands for *M*emory *R*ecall.)

AC Clears the display and erases the number in memory. (AC stands for *A*ll *C*lear.) Press this key before starting a new problem.

EXAMPLE Use a calculator to find the value of $84.50 + (35 × $4.30) − $36.85 + (24 × $5.47).

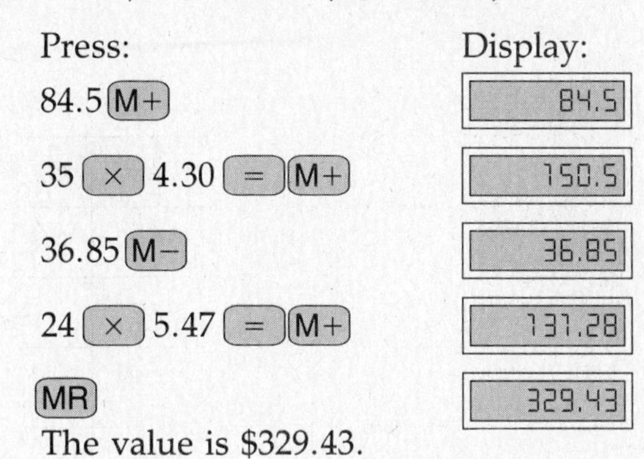

The value is $329.43.

Use a calculator with memory keys. Copy the display for each step and the total. Watch the signs carefully.

									Total value
1.	$65.80	+	(7 × $9.20)	−	($60.76 ÷ 2)	+	$11.98		
	65.8		*64.4*		*30.38*		*11.98*		______
2.	(2 × $5392)	+	$993	−	(16 × $430)	−	$2399		
	______		______		______		______		______
3.	(982 ÷ 2)	−	274	+	(32 × 45)	+	(65 × 30)		
	______		______		______		______		______
4.	(4 × 19.23)	−	(3 × 6.2)	−	2.26	+	(18.4 ÷ 5)		
	______		______		______		______		______
5.	$458.77	+	(3 × $28.95)	+	(2 × $39.44)	+	$12.59		
	______		______		______		______		______

PAYING TAXES

Using Tax Tables

The Internal Revenue Service (IRS) provides tax tables with the instructions for federal income tax forms. If you know a person's taxable income and filing status, you can use the tables to find the tax liability. The columns at the right were taken from the tables provided for 1993, which cover incomes from $0 to $100,000.

EXAMPLE Alex Jones has a taxable income of $17,438. His filing status is single, and $2,750 was withheld from his pay. Does he owe tax or get a refund? How much?

First, find the $17,400–$17,450 income line in the table. Next, find the column for Single. The amount shown where the income line and the status column meet is $2,614. This is the amount of income tax that Alex owes.

Since more than the tax amount was withheld, Alex is due a refund. The refund is $2,750 – $2,614, or $136.

Use the columns from the table to find the tax liabilities for persons with the following taxable incomes and filing status

1. $18,760 Single __________

2. $32,576 Married filing jointly __________

3. $24,765 Head of a household __________

4. $23,770 Single __________

5. $34,834 Married filing separately __________

6. $48,779 Married filing jointly __________

7. Carmen Adams has a taxable income of $48,325. Her filing status is single and $10,500 was withheld from her pay.

   What is her tax liability? __________

   Does she owe or get a refund? How much? __________

8. Jason and Jessie Jones have a taxable income of $33,529. They are married and file a joint return. A total of $5,225 tax was withheld from their pay.

   What is their tax liability? __________

   Do they owe or get a refund and how much? __________

1993 Tax Table—*Continued*

If Form 1040A, line 22, is—		And you are—			
At least	But less than	Single	Married filing jointly *	Married filing sepa-rately	Head of a house-hold
		Your tax is—			
17,000					
17,000	17,050	2,554	2,554	2,554	2,554
17,050	17,100	2,561	2,561	2,561	2,561
17,100	17,150	2,569	2,569	2,569	2,569
17,150	17,200	2,576	2,576	2,576	2,576
17,200	17,250	2,584	2,584	2,584	2,584
17,250	17,300	2,591	2,591	2,591	2,591
17,300	17,350	2,599	2,599	2,599	2,599
17,350	17,400	2,606	2,606	2,606	2,606
17,400	17,450	2,614	2,614	2,614	2,614
17,450	17,500	2,621	2,621	2,621	2,621
17,500	17,550	2,629	2,629	2,629	2,629
17,550	17,600	2,636	2,636	2,636	2,636
17,600	17,650	2,644	2,644	2,644	2,644
17,650	17,700	2,651	2,651	2,651	2,651
17,700	17,750	2,659	2,659	2,659	2,659
17,750	17,800	2,666	2,666	2,666	2,666
17,800	17,850	2,674	2,674	2,674	2,674
17,850	17,900	2,681	2,681	2,681	2,681
17,900	17,950	2,689	2,689	2,689	2,689
17,950	18,000	2,696	2,696	2,696	2,696
18,000					
18,000	18,050	2,704	2,704	2,704	2,704
18,050	18,100	2,711	2,711	2,711	2,711
18,100	18,150	2,719	2,719	2,719	2,719
18,150	18,200	2,726	2,726	2,726	2,726
18,200	18,250	2,734	2,734	2,734	2,734
18,250	18,300	2,741	2,741	2,741	2,741
18,300	18,350	2,749	2,749	2,749	2,749
18,350	18,400	2,756	2,756	2,756	2,756
18,400	18,450	2,764	2,764	2,764	2,764
18,450	18,500	2,771	2,771	2,775	2,771
18,500	18,550	2,779	2,779	2,789	2,779
18,550	18,600	2,786	2,786	2,803	2,786
18,600	18,650	2,794	2,794	2,817	2,794
18,650	18,700	2,801	2,801	2,831	2,801
18,700	18,750	2,809	2,809	2,845	2,809
18,750	18,800	2,816	2,816	2,859	2,816
18,800	18,850	2,824	2,824	2,873	2,824
18,850	18,900	2,831	2,831	2,887	2,831
18,900	18,950	2,839	2,839	2,901	2,839
18,950	19,000	2,846	2,846	2,915	2,846
19,000					
19,000	19,050	2,854	2,854	2,929	2,854
19,050	19,100	2,861	2,861	2,943	2,861
19,100	19,150	2,869	2,869	2,957	2,869
19,150	19,200	2,876	2,876	2,971	2,876
19,200	19,250	2,884	2,884	2,985	2,884
19,250	19,300	2,891	2,891	2,999	2,891
19,300	19,350	2,899	2,899	3,013	2,899
19,350	19,400	2,906	2,906	3,027	2,906
19,400	19,450	2,914	2,914	3,041	2,914
19,450	19,500	2,921	2,921	3,055	2,921
19,500	19,550	2,929	2,929	3,069	2,929
19,550	19,600	2,936	2,936	3,083	2,936
19,600	19,650	2,944	2,944	3,097	2,944
19,650	19,700	2,951	2,951	3,111	2,951
19,700	19,750	2,959	2,959	3,125	2,959
19,750	19,800	2,966	2,966	3,139	2,966
19,800	19,850	2,974	2,974	3,153	2,974
19,850	19,900	2,981	2,981	3,167	2,981
19,900	19,950	2,989	2,989	3,181	2,989
19,950	20,000	2,996	2,996	3,195	2,996

If Form 1040A, line 22, is—		And you are—			
At least	But less than	Single	Married filing jointly *	Married filing separately	Head of a household
		Your tax is—			
23,000					
23,000	**23,050**	3,574	3,454	4,049	3,454
23,050	**23,100**	3,588	3,461	4,063	3,461
23,100	**23,150**	3,602	3,469	4,077	3,469
23,150	**23,200**	3,616	3,476	4,091	3,476
23,200	**23,250**	3,630	3,484	4,105	3,484
23,250	**23,300**	3,644	3,491	4,119	3,491
23,300	**23,350**	3,658	3,499	4,133	3,499
23,350	**23,400**	3,672	3,506	4,147	3,506
23,400	**23,450**	3,686	3,514	4,161	3,514
23,450	**23,500**	3,700	3,521	4,175	3,521
23,500	**23,550**	3,714	3,529	4,189	3,529
23,550	**23,600**	3,728	3,536	4,203	3,536
23,600	**23,650**	3,742	3,544	4,217	3,544
23,650	**23,700**	3,756	3,551	4,231	3,551
23,700	**23,750**	3,770	3,559	4,245	3,559
23,750	**23,800**	3,784	3,566	4,259	3,566
23,800	**23,850**	3,798	3,574	4,273	3,574
23,850	**23,900**	3,812	3,581	4,287	3,581
23,900	**23,950**	3,826	3,589	4,301	3,589
23,950	**24,000**	3,840	3,596	4,315	3,596
24,000					
24,000	**24,050**	3,854	3,604	4,329	3,604
24,050	**24,100**	3,868	3,611	4,343	3,611
24,100	**24,150**	3,882	3,619	4,357	3,619
24,150	**24,200**	3,896	3,626	4,371	3,626
24,200	**24,250**	3,910	3,634	4,385	3,634
24,250	**24,300**	3,924	3,641	4,399	3,641
24,300	**24,350**	3,938	3,649	4,413	3,649
24,350	**24,400**	3,952	3,656	4,427	3,656
24,400	**24,450**	3,966	3,664	4,441	3,664
24,450	**24,500**	3,980	3,671	4,455	3,671
24,500	**24,550**	3,994	3,679	4,469	3,679
24,550	**24,600**	4,008	3,686	4,483	3,686
24,600	**24,650**	4,022	3,694	4,497	3,694
24,650	**24,700**	4,036	3,701	4,511	3,701
24,700	**24,750**	4,050	3,709	4,525	3,709
24,750	**24,800**	4,064	3,716	4,539	3,716
24,800	**24,850**	4,078	3,724	4,553	3,724
24,850	**24,900**	4,092	3,731	4,567	3,731
24,900	**24,950**	4,106	3,739	4,581	3,739
24,950	**25,000**	4,120	3,746	4,595	3,746
25,000					
25,000	**25,050**	4,134	3,754	4,609	3,754
25,050	**25,100**	4,148	3,761	4,623	3,761
25,100	**25,150**	4,162	3,769	4,637	3,769
25,150	**25,200**	4,176	3,776	4,651	3,776
25,200	**25,250**	4,190	3,784	4,665	3,784
25,250	**25,300**	4,204	3,791	4,679	3,791
25,300	**25,350**	4,218	3,799	4,693	3,799
25,350	**25,400**	4,232	3,806	4,707	3,806
25,400	**25,450**	4,246	3,814	4,721	3,814
25,450	**25,500**	4,260	3,821	4,735	3,821
25,500	**25,550**	4,274	3,829	4,749	3,829
25,550	**25,600**	4,288	3,836	4,763	3,836
25,600	**25,650**	4,302	3,844	4,777	3,844
25,650	**25,700**	4,316	3,851	4,791	3,851
25,700	**25,750**	4,330	3,859	4,805	3,859
25,750	**25,800**	4,344	3,866	4,819	3,866
25,800	**25,850**	4,358	3,874	4,833	3,874
25,850	**25,900**	4,372	3,881	4,847	3,881
25,900	**25,950**	4,386	3,889	4,861	3,889
25,950	**26,000**	4,400	3,896	4,875	3,896

If Form 1040A, line 22, is—		And you are—			
At least	But less than	Single	Married filing jointly *	Married filing separately	Head of a household
		Your tax is—			
32,000					
32,000	**32,050**	6,094	4,804	6,569	5,119
32,050	**32,100**	6,108	4,811	6,583	5,133
32,100	**32,150**	6,122	4,819	6,597	5,147
32,150	**32,200**	6,136	4,826	6,611	5,161
32,200	**32,250**	6,150	4,834	6,625	5,175
32,250	**32,300**	6,164	4,841	6,639	5,189
32,300	**32,350**	6,178	4,849	6,653	5,203
32,350	**32,400**	6,192	4,856	6,667	5,217
32,400	**32,450**	6,206	4,864	6,681	5,231
32,450	**32,500**	6,220	4,871	6,695	5,245
32,500	**32,550**	6,234	4,879	6,709	5,259
32,550	**32,600**	6,248	4,886	6,723	5,273
32,600	**32,650**	6,262	4,894	6,737	5,287
32,650	**32,700**	6,276	4,901	6,751	5,301
32,700	**32,750**	6,290	4,909	6,765	5,315
32,750	**32,800**	6,304	4,916	6,779	5,329
32,800	**32,850**	6,318	4,924	6,793	5,343
32,850	**32,900**	6,332	4,931	6,807	5,357
32,900	**32,950**	6,346	4,939	6,821	5,371
32,950	**33,000**	6,360	4,946	6,835	5,385
33,000					
33,000	**33,050**	6,374	4,954	6,849	5,399
33,050	**33,100**	6,388	4,961	6,863	5,413
33,100	**33,150**	6,402	4,969	6,877	5,427
33,150	**33,200**	6,416	4,976	6,891	5,441
33,200	**33,250**	6,430	4,984	6,905	5,455
33,250	**33,300**	6,444	4,991	6,919	5,469
33,300	**33,350**	6,458	4,999	6,933	5,483
33,350	**33,400**	6,472	5,006	6,947	5,497
33,400	**33,450**	6,486	5,014	6,961	5,511
33,450	**33,500**	6,500	5,021	6,975	5,525
33,500	**33,550**	6,514	5,029	6,989	5,539
33,550	**33,600**	6,528	5,036	7,003	5,553
33,600	**33,650**	6,542	5,044	7,017	5,567
33,650	**33,700**	6,556	5,051	7,031	5,581
33,700	**33,750**	6,570	5,059	7,045	5,595
33,750	**33,800**	6,584	5,066	7,059	5,609
33,800	**33,850**	6,598	5,074	7,073	5,623
33,850	**33,900**	6,612	5,081	7,087	5,637
33,900	**33,950**	6,626	5,089	7,101	5,651
33,950	**34,000**	6,640	5,096	7,115	5,665
34,000					
34,000	**34,050**	6,654	5,104	7,129	5,679
34,050	**34,100**	6,668	5,111	7,143	5,693
34,100	**34,150**	6,682	5,119	7,157	5,707
34,150	**34,200**	6,696	5,126	7,171	5,721
34,200	**34,250**	6,710	5,134	7,185	5,735
34,250	**34,300**	6,724	5,141	7,199	5,749
34,300	**34,350**	6,738	5,149	7,213	5,763
34,350	**34,400**	6,752	5,156	7,227	5,777
34,400	**34,450**	6,766	5,164	7,241	5,791
34,450	**34,500**	6,780	5,171	7,255	5,805
34,500	**34,550**	6,794	5,179	7,269	5,819
34,550	**34,600**	6,808	5,186	7,283	5,833
34,600	**34,650**	6,822	5,194	7,297	5,847
34,650	**34,700**	6,836	5,201	7,311	5,861
34,700	**34,750**	6,850	5,209	7,325	5,875
34,750	**34,800**	6,864	5,216	7,339	5,889
34,800	**34,850**	6,878	5,224	7,353	5,903
34,850	**34,900**	6,892	5,231	7,367	5,917
34,900	**34,950**	6,906	5,239	7,381	5,931
34,950	**35,000**	6,920	5,246	7,395	5,945

If Form 1040A, line 22, is—		And you are—			
At least	But less than	Single	Married filing jointly *	Married filing separately	Head of a household
		Your tax is—			
47,000					
47,000	**47,050**	10,294	8,370	10,842	9,319
47,050	**47,100**	10,308	8,384	10,858	9,333
47,100	**47,150**	10,322	8,398	10,873	9,347
47,150	**47,200**	10,336	8,412	10,889	9,361
47,200	**47,250**	10,350	8,426	10,904	9,375
47,250	**47,300**	10,364	8,440	10,920	9,389
47,300	**47,350**	10,378	8,454	10,935	9,403
47,350	**47,400**	10,392	8,468	10,951	9,417
47,400	**47,450**	10,406	8,482	10,966	9,431
47,450	**47,500**	10,420	8,496	10,982	9,445
47,500	**47,550**	10,434	8,510	10,997	9,459
47,550	**47,600**	10,448	8,524	11,013	9,473
47,600	**47,650**	10,462	8,538	11,028	9,487
47,650	**47,700**	10,476	8,552	11,044	9,501
47,700	**47,750**	10,490	8,566	11,059	9,515
47,750	**47,800**	10,504	8,580	11,075	9,529
47,800	**47,850**	10,518	8,594	11,090	9,543
47,850	**47,900**	10,532	8,608	11,106	9,557
47,900	**47,950**	10,546	8,622	11,121	9,571
47,950	**48,000**	10,560	8,636	11,137	9,585
48,000					
48,000	**48,050**	10,574	8,650	11,152	9,599
48,050	**48,100**	10,588	8,664	11,168	9,613
48,100	**48,150**	10,602	8,678	11,183	9,627
48,150	**48,200**	10,616	8,692	11,199	9,641
48,200	**48,250**	10,630	8,706	11,214	9,655
48,250	**48,300**	10,644	8,720	11,230	9,669
48,300	**48,350**	10,658	8,734	11,245	9,683
48,350	**48,400**	10,672	8,748	11,261	9,697
48,400	**48,450**	10,686	8,762	11,276	9,711
48,450	**48,500**	10,700	8,776	11,292	9,725
48,500	**48,550**	10,714	8,790	11,307	9,739
48,550	**48,600**	10,728	8,804	11,323	9,753
48,600	**48,650**	10,742	8,818	11,338	9,767
48,650	**48,700**	10,756	8,832	11,354	9,781
48,700	**48,750**	10,770	8,846	11,369	9,795
48,750	**48,800**	10,784	8,860	11,385	9,809
48,800	**48,850**	10,798	8,874	11,400	9,823
48,850	**48,900**	10,812	8,888	11,416	9,837
48,900	**48,950**	10,826	8,902	11,431	9,851
48,950	**49,000**	10,840	8,916	11,447	9,865
49,000					
49,000	**49,050**	10,854	8,930	11,462	9,879
49,050	**49,100**	10,868	8,944	11,478	9,893
49,100	**49,150**	10,882	8,958	11,493	9,907
49,150	**49,200**	10,896	8,972	11,509	9,921
49,200	**49,250**	10,910	8,986	11,524	9,935
49,250	**49,300**	10,924	9,000	11,540	9,949
49,300	**49,350**	10,938	9,014	11,555	9,963
49,350	**49,400**	10,952	9,028	11,571	9,977
49,400	**49,450**	10,966	9,042	11,586	9,991
49,450	**49,500**	10,980	9,056	11,602	10,005
49,500	**49,550**	10,994	9,070	11,617	10,019
49,550	**49,600**	11,008	9,084	11,633	10,033
49,600	**49,650**	11,022	9,098	11,648	10,047
49,650	**49,700**	11,036	9,112	11,664	10,061
49,700	**49,750**	11,050	9,126	11,679	10,075
49,750	**49,800**	11,064	9,140	11,695	10,089
49,800	**49,850**	11,078	9,154	11,710	10,103
49,850	**49,900**	11,092	9,168	11,726	10,117
49,900	**49,950**	11,106	9,182	11,741	10,131
49,950	**50,000**	11,120	9,196	11,757	10,145

50,000 or over — use Form 1040

* This column must also be used by a qualifying widow(er).

PROBLEM-SOLVING STRATEGY

Use Guess-and-Check

A good way to solve some problems is to guess the answer. Then check to see if the answer is correct, and make another guess if necessary. Your next guess will be better because you will learn from the first guess. Continue to guess and check until you find the correct answer.

Read the problem.

In one year, Mr. and Mrs. Yates had earnings of $40,600. If Mr. Yates earned $2200 more than Mrs. Yates, how much did each person earn?

Guess and check.

For Mr. Yates, start with a guess that is more than half of $40,600. Subtract to find Mrs. Yates's earnings. Then check whether the difference is $2200.

Guess 1:		*Check:*
Mr. Yates	$21,000	The total earnings
Mrs. Yates	$19,600	are $40,600 but the
Difference	$ 1,400	difference is less than $2200.

Guess 2:		*Check:*
Mr. Yates	$21,500	The total earnings
Mrs. Yates	$19,100	are $40,600 but the
Difference	$ 2,400	difference is more than $2200.

Guess 3:		*Check:*
Mr. Yates	$21,400	Both the total
Mrs. Yates	$19,200	earnings and the
Difference	$ 2,200	difference are correct.

So, Mr. Yates earned $21,400 and Mrs. Yates earned $19,200.

Solve. Use the guess-and-check strategy.

1. Linda received interest income of $375 from two accounts. One account earned $35 more than the other. How much interest income did Linda receive from each account?

Answer ______________

2. Jason earned a total of $15,900 from two part-time jobs. If he earned $3700 more from one job than the other, how much did he earn from each job?

Answer ______________

Solve. Use the guess-and-check strategy.

3. Mr. and Mrs. King earned a total of $50,700. Mrs. King earned twice as much as Mr. King. How much did each person earn?

Answer ______________________

4. Mr. and Mrs. Rosenbaum received interest income of $924 from two accounts. One account earned 3 times as much as the other. How much interest income did they receive from each account?

Answer ______________________

5. On one of Nellie's paychecks, $85 was withheld for state and federal taxes. The amount withheld for federal tax was $53 more than the amount withheld for state tax. How much was withheld for federal tax?

Answer ______________________

6. A total of $140 was deducted from Wally's paycheck for taxes and insurance. The deduction for insurance was $10 less than the deduction for taxes. How much was deducted for insurance?

Answer ______________________

7. In three weeks, a total of $191 was withheld from Teresa's paycheck for federal tax. The same amount was withheld for Weeks 1 and 2. The amount withheld for Week 3 was $5 more. How much was withheld in Week 3?

Answer ______________________

8. Brent earned $35 for three baby-sitting jobs. He earned $5 more from Job A than from Job B. He earned half as much from Job C as he did from Job B. How much did he earn from each job?

Answer ______________________

Form 1040A: Adjusted Gross Income

Form 1040A can be used by single or married persons with or without dependents. A person who does not itemize or does not need to file special forms can use 1040A. The first page of Form 1040A is shown on the page at the right.

Label

Taxpayers must write or attach a label that shows their name, address, and social security number. They can also check a box to indicate that they want $3 to go to the Presidential Campaign Fund.

Check your filing status

Taxpayers should check only one box.

- *Single* is for a person who is unmarried or widowed.
- *Married filing joint return* is for married persons who combine their incomes to file a single return, even if only one had an income.
- *Married filing separate return* is for married persons who file individually. Generally, this results in the two paying more tax than if they file a joint return.
- *Head of household* is for an unmarried person who provides a home for another.
- *Qualifying widow(er) with dependent child* is for a person whose spouse died within the two years prior to the tax year.

Figure your exemptions

Taxpayers list their dependents. For each exemption, including the taxpayer, $2,350 can be deducted on line 21.

Figure your total income

Taxpayers must declare all income, including salaries, tips, tax refunds, bank interest, and stock dividends. If the total interest and dividends are greater than $400, then a separate form must be attached to Form 1040A.

Figure your adjusted gross income

Taxpayers can deduct qualified contributions to an **Individual Retirement Account (IRA)** from their total income to find their adjusted gross income.

Follow these instructions to complete the first page of Form 1040A. Use the form at the right.

Complete the **label** section.

1. Write the names Luis J. and Eva M. Sanchez. Give them an address and social security numbers. They both want $3 to go to the Presidential Election Campaign Fund.

Complete the **filing-status** section.

2. Check the box for Luis and Eva's filing status. They are filing a joint return.

Complete the **exemptions** section.

3. Show that Luis and Eva have 3 dependent children: Carlos, Gloria, and Ramon. All of the children are older than age 5, younger than age 13, and have lived at home all year. Give each child a social security number.

Complete the **total income section.**

4. Luis's W-2 showed that he earned $20,530 in wages. Eva's W-2 showed that she earned $24,188. Write the total earnings on line 7.

5. The Sanchez's joint savings account earned $263 interest, and they earned $130 interest on tax-free city savings bonds. They also earned stock dividends of $301. Write these amounts on lines 8a, 8b, and 9. Leave lines 10a through 13b blank.

6. Follow the instructions on line 14 to find the total income for Luis and Eva Sanchez.

Complete the **adjusted gross income** section.

7. Luis and Eva Sanchez can each deduct $2000 for IRA contributions. Complete each part of line 15 to show this.

8. Follow the directions given in line 16 to find the adjusted gross income for Luis and Eva Sanchez.

You may wish to make copies of this tax form to use for exercises on other pages. This form is a sample only. Because tax rules change frequently, new tax forms are distributed each year by the IRS.

Form **1040A** Department of the Treasury—Internal Revenue Service

U.S. Individual Income Tax Return (U) **19—** IRS Use Only—Do not write or staple in this space.

OMB No. 1545-0085

Label (See page 15.)

Use the IRS label. Otherwise, please print or type.

LABEL HERE

Your first name and initial | Last name

If a joint return, spouse's first name and initial | Last name

Home address (number and street). If you have a P.O. box, see page 16. | Apt. no.

City, town or post office, state, and ZIP code. If you have a foreign address, see page 16.

Your social security number

Spouse's social security number

For Privacy Act and Paperwork Reduction Act Notice, see page 4.

Presidential Election Campaign Fund (See page 16.) **Yes** **No**

Do you want $3 to go to this fund? . . . .

If a joint return, does your spouse want $3 to go to this fund?

Note: *Checking "Yes" will not change your tax or reduce your refund.*

Check the box for your filing status

(See page 16.)

Check only one box.

1 ☐ Single
2 ☐ Married filing joint return (even if only one had income)
3 ☐ Married filing separate return. Enter spouse's social security number above and full name here. ▶
4 ☐ Head of household (with qualifying person). (See page 17.) If the qualifying person is a child but not your dependent, enter this child's name here. ▶
5 ☐ Qualifying widow(er) with dependent child (year spouse died ▶ 19). (See page 18.)

Figure your exemptions

(See page 19.)

If more than seven dependents, see page 22.

6a ☐ **Yourself.** If your parent (or someone else) can claim you as a dependent on his or her tax return, **do not** check box 6a. But be sure to check the box on line 18b on page 2.

b ☐ **Spouse**

No. of boxes checked on 6a and 6b ____

c **Dependents:**

(1) Name (first, initial, and last name)	(2) Check if under age 1	(3) If age 1 or older, dependent's social security number	(4) Dependent's relationship to you	(5) No. of months lived in your home in 1993

No. of your children on 6c who:

• **lived with you** ____

• **didn't live with you due to divorce or separation (see page 22)** ____

Dependents on 6c not entered above ____

d If your child didn't live with you but is claimed as your dependent under a pre-1985 agreement, check here . . . . ▶ ☐

e Total number of exemptions claimed.

Add numbers entered on lines above ☐

Figure your total income

Attach Copy B of your Forms W-2 and 1099-R here.

If you didn't get a W-2, see page 24.

If you are attaching a check or money order, put it on top of any Forms W-2 or 1099-R.

7 Wages, salaries, tips, etc. This should be shown in box 1 of your W-2 form(s). Attach Form(s) W-2. **7**

8a **Taxable** interest income (see page 25). If over $400, also complete and attach Schedule 1, Part I. **8a**

b **Tax-exempt** interest. DO NOT include on line 8a. **8b**

9 Dividends. If over $400, also complete and attach Schedule 1, Part II. **9**

10a Total IRA distributions. **10a** | **10b** Taxable amount (see page 26). **10b**

11a Total pensions and annuities. **11a** | **11b** Taxable amount (see page 26). **11b**

12 Unemployment compensation (see page 30). **12**

13a Social security benefits. **13a** | **13b** Taxable amount (see page 30). **13b**

14 Add lines 7 through 13b (far right column). This is your **total income.** ▶ **14**

Figure your adjusted gross income

15a Your IRA deduction (see page 32). **15a**

b Spouse's IRA deduction (see page 32). **15b**

c Add lines 15a and 15b. These are your **total adjustments.** **15c**

16 Subtract line 15c from line 14. This is your **adjusted gross income.** If less than $23,050 and a child lived with you, see page 63 to find out if you can claim the "Earned income credit" on line 28c. ▶ **16**

Form 1040A page 1

Form 1040A: Taxable Income and Credits

The second page of Form 1040A is shown on the page at the right.

Figure your standard deduction, exemption amount, and taxable income
For most persons, the **standard deduction** is determined by the their filing status as indicated by a check mark on line 1, 2, 3, 4, or 5 on the first page of the form. For such persons, the amount of the deduction can be found on line 19.

The standard deduction is greater for taxpayers who are older than 65 and/or are blind. Also, the deduction may be less for a single person who earns interest or has other income and is claimed as a dependent on another person's return. An example of this is a young child with a savings account or a teenager with a part-time job.

On line 21 the total number of exemptions listed on line 6e of page 1 are multiplied by $2,300 and used to find the **taxable income.**

Figure your tax, credits, and payments
Most taxpayers find their tax using a tax table. Samples from this table are shown on pages 156 and 157.

Note that line 24a allows **credit for child and dependent care expenses** that enable the taxpayer to work. To claim such credits, the taxpayer must complete another form, called Schedule 2, and attach it to this form. The credit ranges from 20 to 30% of qualified expenses.

Line 28c allows an earned income credit for certain persons who have an adjusted gross income of less than $23,050 and have at least one dependent child. This credit can be as much as $1,434 for one qualifying child and $1,511 for two or more qualifying children.

Figure your refund or amount you owe
Taxpayers compute the amount owed the government or the amount due as a refund.

Sign your return
Taxpayers sign the form, write the date, and list their occupations.

Paid preparer's use only
Anyone other than the taxpayer who has prepared the form and has charged for its preparation must sign the form.

Follow these instructions to complete the second page of Form 1040A for Luis and Eva Sanchez. Use the form at the right.

Complete the **taxable income** section.

1. Copy the amount from line 16 on page 1 of the form to line 17 at the top of page 2. Lines 18a, 18b, and 18c do not apply to the Sanchez family. Use the information given with line 19 to enter the standard deduction for Luis and Eva Sanchez.

2. Multiply the total number of exemptions claimed on line 6e on page 1 by $2,350. Write the product on line 21. Subtract to find the **taxable income** and write it on line 22.

Complete the **tax, credits, and payments** section.

3. Use the tax tables on pages 156–157 to find the tax on the amount shown on line 22. Write the answer on line 23.

4. The Sanchez family can get a child care credit of 20% of $4800, or $960. Write this amount on lines 24a and 24c. Subtract line 24c from line 23 to find your total tax. Write it on lines 25 and 27.

5. Luis and Eva Sanchez had two W-2 forms, and the amounts that were withheld for federal tax were $1460.24 and $1552.83. Write the total federal tax withheld, to the nearest dollar, on lines 28a and 28d.

Complete the **refund or amount you owe** section.

6. Compare the amounts on lines 27 and 28d. If line 28d is greater, write the difference on line 29. This is the refund. If line 27 is greater, write the difference on line 32. This is the amount owed.

Complete the **sign your return** section

7. The date is April 10, 1994. Luis Sanchez is a security guard and Eva Sanchez is a teacher.

You may wish to make copies of this tax form to use for exercises on other pages. This form is a sample only. Because tax rules change frequently, new tax forms are distributed each year by the IRS.

19— Form **1040A** **Page 2**

Name(s) shown on page 1 | **Your social security number**

Figure your standard deduction, exemption amount, and taxable income

17 Enter the amount from line 16. 17

18a Check if: ☐ **You** were 65 or older ☐ Blind ☐ **Spouse** was 65 or older ☐ Blind **Enter number of boxes checked** ▶ 18a ☐

b If your parent (or someone else) can claim you as a dependent, check here . . . . . ▶ 18b ☐

c If you are married filing separately and your spouse files Form 1040 and itemizes deductions, see page 36 and check here ▶ 18c ☐

19 Enter the **standard deduction** shown below for your filing status. **But if you checked any box on line 18a or b,** go to page 36 to find your standard deduction. **If you checked box 18c,** enter -0-.
- Single—$3,700
- Head of household—$5,450
- Married filing jointly or Qualifying widow(er)—$6,200
- Married filing separately—$3,100

19

20 Subtract line 19 from line 17. If line 19 is more than line 17, enter -0-. 20

21 Multiply $2,350 by the total number of exemptions claimed on line 6e. 21

22 Subtract line 21 from line 20. If line 21 is more than line 20, enter -0-. This is your **taxable income.** ▶ 22

Figure your tax, credits, and payments

If you want the IRS to figure your tax, see the instructions for line 22 on page 37.

23 Find the tax on the amount on line 22. Check if from: ☐ Tax Table (pages 50–55) or ☐ Form 8615 (see page 38). 23

24a Credit for child and dependent care expenses. Complete and attach Schedule 2. 24a

b Credit for the elderly or the disabled. Complete and attach Schedule 3. 24b

c Add lines 24a and 24b. These are your **total credits.** 24c

25 Subtract line 24c from line 23. If line 24c is more than line 23, enter -0-. 25

26 Advance earned income credit payments from Form W-2. 26

27 Add lines 25 and 26. This is your **total tax.** ▶ 27

28a Total Federal income tax withheld. If any tax is from Form(s) 1099, check here. ▶ ☐ 28a

b 1993 estimated tax payments and amount applied from 1992 return. 28b

c **Earned income credit.** Complete and attach Schedule EIC. 28c

d Add lines 28a, 28b, and 28c. These are your **total payments.** ▶ 28d

Figure your refund or amount you owe

29 If line 28d is more than line 27, subtract line 27 from line 28d. This is the amount you **overpaid.** 29

30 Amount of line 29 you want **refunded to you.** 30

31 Amount of line 29 you want **applied to your 1994 estimated tax.** 31

32 If line 27 is more than line 28d, subtract line 28d from line 27. This is the **amount you owe.** For details on how to pay, including what to write on your payment, see page 42. 32

33 Estimated tax penalty (see page 43). Also, include on line 32. 33

Sign your return

Keep a copy of this return for your records.

Under penalties of perjury, I declare that I have examined this return and accompanying schedules and statements, and to the best of my knowledge and belief, they are true, correct, and accurately list all amounts and sources of income I received during the tax year. Declaration of preparer (other than the taxpayer) is based on all information of which the preparer has any knowledge.

Your signature	Date	Your occupation
Spouse's signature. If joint return, BOTH must sign.	Date	Spouse's occupation

Paid preparer's use only

Preparer's signature	Date	Check if self-employed ☐	Preparer's social security no.
Firm's name (or yours if self-employed) and address		E.I. No.	
		ZIP code	

Form 1040A page 2

Practice with Form 1040A

Complete each column of the chart. In each column, either line 29 or line 30 should be blank. Refer to a copy of Form 1040A.

		1.	2.	3.	4.	5.
NAME OF TAXPAYER		**Bernard**	**Nguyen**	**Robinson**	**Montoya**	**Kanui**
FILING STATUS		**Married filing jointly**	**Single**	**Head of household**	**Married filing jointly**	**Married filing sep.**
NUMBER OF EXEMPTIONS		4	1	3	2	1
Line 17	Adj. gross income	$40,185	$24,480	$46,460	$58,580	$29,410
Line 19	Standard deduction	6,200	3,700	5,450	6,200	3,100
Line 20	Line 18 – line 19	*33,985*				
Line 21	Multiply $2350 by number of exemptions	*9,400*				
Line 22	Taxable income	*24,585*				
Line 23	Tax from table (See pages 156–157.)	*3,686*				
Line 24a	Credit for child care	1,080	0	704	0	0
Line 27	Total tax (Line 23 – line 24a)	*2,606*				
Line 28d	Tax withheld plus earned income credit	2,738	2,645	5,550	7,258	3,580
Line 29	Refund	*132*				
Line 30	Amount owed					

6. Mr. and Mrs. Hannah had a total income of $42,634. They could deduct $4000 for IRA contributions. They claimed 4 exemptions and took the standard deduction of $6200. What was their taxable income?

Answer ____________________

7. Use the table on pages 156–157 to find Mr. and Mrs. Hannah's income tax. They can subtract a credit of 20% of $2370 for child care. What is the tax liability of Mr. and Mrs. Hannah?

Answer ____________________

8. Tina Storment is filing as head of household. She qualifies for an earned income credit of $320. Her employer withheld $1,354 of federal tax. If the tax on her income is $1,504, does she owe or get a refund? How much?

Answer ____________________

9. Kasia Smith's W-2 showed earnings of $28,967. She also received $783 of interest income, but $125 of this was tax-exempt. What was Kasia's total taxable income?

Answer ____________________

Real Estate Taxes

If you own a home or other property, you must pay **real estate** taxes to the city and/or county government. The money collected is used for schools and other local government services. The amount of real estate tax depends on the assessed value of the property and the tax rate for the location. The **assessed value** is found by multiplying the market value of the property by the rate of assessment (a percent). The market value is determined by the county assessor's office. The tax rate is determined by the city and/or county government.

The tax rate is often expressed in mills per dollar of assessed value. A **mill** is $\frac{1}{10}$ cent or $\frac{1}{1000}$ of a dollar. A yearly tax rate of 45 mills per dollar is equal to $45 per $1000 of assessed value.

EXAMPLE The market value of a home is $75,400. The rate of assessment is 40% and the yearly tax rate is 52 mills. What is the assessed value of the property? How much real estate tax is owed for a year?

Assessed value: 40% of $75,400 = $30,160
Tax rate: 52 mills or $52 per $1000
Assessed value in thousands: 30.16 (30,160 ÷ 1000)
Real estate tax: $52 × 30.16 = $1568.32

If a home is mortgaged the lender may collect a portion of the real estate tax with each mortgage payment. The lender keeps the taxes in a special account called an **escrow** account, and pays each real estate tax bill when it is due. For the home described above, the lender would collect $\frac{1}{12}$ of $1568.32 each month. This is about $131 per month.

Complete the chart to find the yearly and monthly real estate tax for each home.

	Market Value	Rate of Assessment	Assessed Value	Assessed Value ÷ $1000	Tax Rate in Mills	Yearly Real Estate Tax	Monthly Real Estate Tax
1.	$58,000	30%	*$17,400*	*17.4*	46		
2.	$79,000	40%			55		
3.	$48,000	35%			38		
4.	$97,000	45%			27		
5.	$65,000	35%			16		
6.	$135,000	45%			40		
7.	$14,000	30%			51		
8.	$199,000	40%			48		

Itemized Deductions for Form 1040

Most taxpayers can subtract a standard deduction from their income, based on their filing status. For example, looking at line 19 of Tax Form 1040A on page 163, you see that the standard deduction for a taxpayer filing as a single person is $3700. Some persons, especially those with home mortgages, may pay less tax by **itemizing deductions** instead of taking the standard deduction. Taxpayers who decide to itemize must use Form 1040 and complete Schedule A: Itemized Deductions.

Expenses that may be itemized include home mortgage interest, real estate taxes, state and local income taxes, state sales taxes, and gifts to charity. Also, if a person has very high medical expenses or casualty and theft losses, a portion of these expenses may be deductible. To decide whether to itemize deductions, add the deductible expenses. If the total is more that the standard deduction, it would be worthwhile to itemize.

EXAMPLE Mr. and Mrs. Draper are filing a joint income tax return. They have a bank statement showing that they paid $3672 in home mortgage interest and $1340 in real estate taxes. Their W-2 forms show that they paid $1410 in state and local income taxes. Also, they have receipts for gifts of $1300 to charity. Should they itemize their deductions?

$$\$3672 + \$1340 + \$1410 + \$1300 = \$7722$$

Mr. and Mrs. Draper should itemize since $7722 is more than $6200, the standard deduction for a married couple filing jointly.

	Name	Standard Deduction	Itemized Deductions				Total	Should the person itemize?
			Mortgage Int.	Taxes	Charity	Other		
1.	A. Burns	$5450	$3178	$739	$745	0		
2.	W. West	$3700	$3102	$2650	$936	$420		
3.	R. Santos	$6200	0	$3877	$1493	$138		
4.	S. Cohen	$3700	0	$1100	$960	$307		
5.	C. Cheng	$6200	$5205	$4111	$4528	0		
6.	P. Kelly	$3100	$2190	$1784	$2310	$673		

7. In one year, Morris paid $2483 in medical expenses. His adjusted gross income was $21,380. He can deduct the portion of the medical expenses that exceeds 7.5% of $21,380. How much can he deduct for medical expenses? (Hint: Find $0.075 \times \$21,380$ and subtract the result from $2483.)

Answer ____________________

Form 1040

Form 1040, called the long form, must be used by persons who itemize deductions on Schedule A. It must also be used by persons who have income from self-employment, farming, rental property, or certain other sources.

Persons who own a business or have self-employment income must complete Schedule C, *Profit or Loss from Business,* in addition to Form 1040. If the profit is over $400, they must also complete Schedule SE, *Self-Employment Tax.* The self-employment tax rate is 15.3% of net earnings. This rate is a total of 12.4% for social security and 2.9% for Medicare.

EXAMPLE Paula Albright is self-employed. She made a profit of $38,475. Only $33,475 of this was taxable income. Her filing status is single. What is Paula's total tax liability?

Use the table on pages 156–157 to find the amount of income tax she must pay on $33,475. .... $6,500
Then compute the self-employment tax by finding 15.3% of $38,475 to the nearest whole dollar. $5,887
Find the total. $12,387
Paula's total tax liability is $12,387.

Solve.

1. Ricardo's business income from Schedule C was $8075. His income from interest and dividends was $723, and he also had wages of $12,872. What was his total income?

Answer ____________________

2. Ricardo must pay $3251 tax on his income. He also must pay 15.3% self-employment tax on his business income? What is Ricardo's total tax liability, rounded to the nearest dollar?

Answer ____________________

3. Maria's filing status is single. Her adjusted gross income for a year was $32,870. Find her taxable income by subtracting her itemized deductions, $5680, and one exemption, $2350.

Answer ____________________

4. Find Maria's tax liability by using the tables on pages 156–157. Maria has no tax credits or other taxes. If $5,320 was withheld, does Maria owe or get a refund? How much?

Answer ____________________

5. Kevin's filing status is head of household. His adjusted gross income for a year was $46,015. Find his taxable income by subtracting his itemized deductions, $6290, and 3 exemptions of $2350 each.

Answer ____________________

6. Kevin can take a tax credit of $770 for child care expenses. Use the tables on pages 156–157 to find the tax on Kevin's taxable income. Then subtract the credit to find his tax liability. If $4448 was withheld, does Kevin owe or get a refund? How much?

Answer ____________________

PROBLEM-SOLVING STRATEGY

Select a Strategy

In this book, you have worked with several different problem-solving strategies. Some of them are listed in the box at the right.

PROBLEM-SOLVING STRATEGIES
- Choose an Operation
- Make a Table
- Make a List
- Use Estimation
- Work Backwards
- Find a Pattern
- Make a Drawing
- Use Guess-and-Check

Read the problem. Select a strategy from the box. Then solve.

1. In one year, Kingo Takata paid a total of $8400 for mortgage interest and real estate taxes. The mortgage interest was $3600 more than the taxes. How much mortgage interest did Kingo pay?

Strategy ____________________

Answer ____________________

2. Patty paid a total of $600 to an accountant who prepared tax forms for her business. Patty was charged $60 for the first hour plus $40 for each additional hour. How many hours did the job take?

Strategy ____________________

Answer ____________________

3. Manny Lincoln's W-2 form showed that his gross income for a year was $26,592. Suppose he earned the same amount each month. What was Manny's gross income per month?

Strategy ____________________

Answer ____________________

4. The checks in Vera's checkbook are numbered from 307 to 325. How many of the check numbers have digits with a sum less than 10? (Hint: The digits of 313 have a sum of 3 + 1 + 3 = 7.)

Strategy ____________________

Answer ____________________

5. Mr. and Mrs. Zeller are filling out Schedule A, *Itemized Deductions.* They can deduct $3250 for taxes and $2500 for contributions to charity. They are also deducting mortgage interest. If their total deductions equal $10,950, how much mortgage interest are Mr. and Mrs. Zeller deducting?

Strategy ____________________

Answer ____________________

6. A large room in a tax preparation office is being separated into 8 foot square work areas. How many work areas can be made from a 16 ft by 24 ft space?

Strategy ____________________

Answer ____________________

Read the problem. Select a strategy from the box. Then solve.

7. A tax preparer charges $8 to fill out a 1040EZ and $15 to fill out a 1040A. In one evening, the tax preparer received $54 for completing some 1040EZ and 1040A forms. How many 1040EZ forms were completed?

Strategy ____________________

Answer ____________________

8. Raymond has four bills to pay, the phone, electric, rent, and a car payment. He is deciding the order in which to pay the bills. How many different orders are possible?

Strategy ____________________

Answer ____________________

9. Mabel must pay about 15% tax on her taxable income of $9812. About how much tax must she pay?

Strategy ____________________

Answer ____________________

10. Mr. and Mrs. Salefski spent a total of $4800 for childcare for their two girls. Childcare expenses for Jenny were twice as much as for Anna. How much was spent for childcare for Anna?

Strategy ____________________

Answer ____________________

11. Carola Delmar received three W-2 forms. The amounts withheld for federal tax were $1152, $294, and $487. What was the total federal income tax withheld?

Strategy ____________________

Answer ____________________

12. George Stefan made profits of $5824 from self-employment. He must pay 15.3% social security tax. About how much social security tax does he owe?

Strategy ____________________

Answer ____________________

13. There is a typing error in the section of the tax table shown below. Which amount is wrong? What might the correct amount be?

Taxable Income	Tax Amount (Single)
29,000-29,049	5,807
29,050-29,099	5,821
29,100-29,149	5,835
29,150-29,199	5,849
29,200-29,249	5,853

Strategy ____________________

Answer ____________________

14. There is a typing error in the section of the tax table shown below. Which amount is wrong? What might the correct amount be?

Taxable Income	Tax Amount (Head of Household)
7,400-7,449	1,114
7,450-7,499	1,112
7,500-7,549	1,129
7,550-5,599	1,136
7,600-7,649	1,144

Strategy ____________________

Answer ____________________

Sales Tax

Most state governments and some local governments collect sales tax on purchases. The tax rate varies from state to state. The sales tax money is used to pay for the cost of government services.

EXAMPLE Enrico buys a pizza for \$8.69. If the sales tax rate is 6%, how much tax does he pay? What is the total cost?

6% of \$8.69 = 0.06 × \$8.69
= \$0.5214 or \$0.53 (Note: Round up to the next cent.)

Total cost = \$8.69 + \$0.53 = \$9.22

He pays 53¢ tax. The total cost is \$9.22.

In some places, the rate of sales tax is lower on food and clothing. States also have different rules about what is taxable. For example, delivery fees and labor charges are taxed in some states but not others.

Find the sales tax and total cost for an item with the given price.

	Price	Sales Tax Rate	Sales Tax	Total Cost
1.	\$200	6%		
2.	\$12.98	8%		
3.	\$88.45	7%		
4.	\$9.73	9%		
5.	\$37.82	6.5%		
6.	\$284.35	8%		
7.	\$10,850	7.5%		
8.	\$0.86	8.5%		

9. Suppose you are buying three items with prices of \$5.75, \$15.34, and \$42.99. The sales tax rate is 6%. How much sales tax will you pay?

Answer ________________

10. A group of people went out to dinner. The bill was \$76.35 plus 5.5% tax. They left a \$12 tip. How much was paid in all? (Hint: Do not include sales tax on the tip.)

Answer ________________

Excise Tax

Federal, state, and local governments may collect additional taxes, called **excise taxes,** on gasoline, cigarettes, liquor, and other specific items. These taxes may not be obvious to the consumer since some are paid by the manufacturer or supplier, but are passed on to the consumer as part of the purchase price.

Over 25% of the amount a customer pays for gasoline is excise tax. Gasoline taxes are collected by the federal government (18.4 cents per gallon), the states (from 7 cents to 20 cents per gallon), and local governments (amount varies). The money collected for gasoline taxes is used to build and repair streets and highways.

EXAMPLE A station charges $1.18 per gallon for gasoline. Of that, 39¢ is excise tax. What percent of the price is excise tax?

39¢ is what percent of $1.18? Think of $1.18 as 118¢.
Divide 39 by 118. $39 \div 118 = 0.3305$
Write the decimal as a percent. 33%

Solve.

1. Suppose you buy 18 gallons of gasoline for $1.32 per gallon. What is your total cost? If the excise tax is 27 cents per gallon, how much excise tax is collected?

Answer ____________

2. Suppose you buy 22 gallons of gasoline in a city that collects excise tax of 17¢ per gallon. If you pay $1.27 per gallon, what percent of the price is city excise tax?

Answer ____________

3. Suppose you live in a state that collects 17.4¢ excise tax on a gallon of gasoline. If you buy 17.6 gallons of gasoline, how much state excise tax is collected?

Answer ____________

4. Leroy buys 24.3 gallons of gasoline for $28.46. The price includes $8.25 for excise tax. What percent of the price is excise tax?

Answer ____________

5. Theresa lives in a state that collects 14.5¢ excise tax per gallon of gasoline. If she buys 19.3 gallons of gasoline, how much state excise tax is collected?

Answer ____________

6. A customer paid $29.86 for 23.7 gallons of gasoline. The excise tax is 32.8¢ a gallon. How much would a gallon of gasoline cost without the excise tax?

Answer ____________

PAYING TAXES

Unit 7 Review

Solve.

1. Brent's monthly salary is $2,147. How much is withheld from Brent's pay each month for social security tax? Use a social security tax rate of 7.65%.

Answer ____________________

2. Walter's W-2 form showed that he earned $14,875 in wages. Of this, $297.50 was withheld for local tax. What percent of his wages were withheld for local tax?

Answer ____________________

3. Vernon is filing form 1040EZ. His adjusted gross income was $22,785. He takes a $3000 standard deduction and a $2000 personal exemption. What is his taxable income? If the tax rate is 15%, what is Vernon's tax liability?

Answer ____________________

4. Mr. and Mrs. Wheetley file a joint tax return using form 1040A. Their adjusted gross income is $45,308, and they take the $6200 standard deduction for married couples. If they claim 4 exemptions at $2350 each, what is their taxable income?

Answer ____________________

5. During the past year, $4381 of Ellen's pay was withheld for federal income tax. If her tax liability is $4628, will she owe or get a refund? How much?

Answer ____________________

6. The market value of a home is $48,000 and the rate of assessment is 45%. The tax rate is 52 mills ($52 per $1000). What is the yearly real estate tax?

Answer ____________________

7. Winona's filing status is head of household. She can deduct $4,359 in mortgage interest and real estate tax, $1,650 for charitable contributions, and $854 in state and local taxes. What is the total of these deductions? Should Winona itemize or take the standard deduction?

Answer ____________________

8. Nancy is self-employed and made a profit of $22,150 last year. She must pay $2572 in income tax. She also must pay social security tax of 13.3% of her profit. What is Nancy's total tax liability? Round to the nearest dollar.

Answer ____________________

9. Monroe bought three items of clothing costing $75.95, $28.65, and $15.88. The sales tax rate was 6.5%. How much sales tax did he pay on the three items? What was the total cost?

Answer ____________________

10. Kendra bought 16 gallons of gasoline for $1.34 per gallon. The price of each gallon included 29.3 cents for gasoline tax. What percent of the price was tax?

Answer ____________________

FINAL REVIEW

Find each answer.

	a	b	c	d	e
1.	$3908 + 765$	$\$65.74 + 3.86$	$2 + 0.99$	$403 + 6795 + 8579$	$0.03 + 3.33 + 33.03$
2.	$42{,}833 - 1{,}976$	$\$30.07 - 12.50$	$16 - 4.783$	$20{,}000 - 999$	$7064 - 195$
3.	42×12	1.64×6	$\$7.49 \times 15$	8.8×0.9	$\$19.99 \times 23$
4.	$24\overline{)1680}$	$8\overline{)\$56.32}$	$4\overline{)3.21}$	$1.2\overline{)66.48}$	$0.75\overline{)2}$

Change each measurement.

	a	b	c
5.	45 m = ______ cm	651 kg = ______ g	3.8 L = ______ mL
6.	6 lb = ______ oz	52 in. = ___ ft ___ in.	15 pt = ______ c

Find each answer. Simplify.

	a	b	c	d
7.	$4 \times \frac{5}{12}$	$\frac{5}{6} \times \frac{1}{5}$	$4\frac{1}{2} \times \frac{2}{9}$	$\frac{1}{3} \times \frac{3}{16}$
8.	$\frac{1}{3} \div 15$	$\frac{7}{8} \div \frac{3}{4}$	$1\frac{1}{4} \div \frac{5}{16}$	$6\frac{1}{9} \div 1\frac{3}{8}$
9.	$\frac{2}{3} + \frac{3}{10}$	$1\frac{3}{4} + 4\frac{2}{5}$	$5 + 2\frac{3}{10}$	$3\frac{3}{16} + \frac{5}{8}$
10.	$\frac{5}{8} - \frac{1}{3}$	$6 - \frac{5}{9}$	$11 - 7\frac{1}{4}$	$6\frac{1}{9} - 2\frac{1}{8}$

Find each number.

	a	*b*
11.	50% of 964	What percent of 1400 is 616?
12.	5% of what number is 45?	99% of 100

Solve.

13. Gary earns $210.40 for a 40-hour workweek. He earns $7.89 for each hour of overtime. How much does he earn for a 48-hour week?

Answer ____________________

14. Allissa is a computer technician who earns a monthly salary of $2750. What is her yearly salary?

Answer ____________________

15. Freddy sells office supplies. He earns a salary of $216 per work plus commission of 6% on all sales over $2000. One week he had sales of $6200. How much did he earn that week?

Answer ____________________

16. In one pay period, Jennifer had $624.80 in regular earnings. Her employer withheld $147.20 for taxes and $65.30 for insurance. What was Jennifer's take-home pay for that pay period?

Answer ____________________

17. Holly's disability insurance pays 60% of her salary for up to 52 weeks. If her weekly salary is $650 and she is disabled for 60 weeks, how much disability pay will she get?

Answer ____________________

18. A store owner bought lamps for $15.70 each and sold them for $27.95. The overhead was $10.80 per lamp. What was the profit on each lamp?

Answer ____________________

19. Pedro has a yearly income of $23,000. He budgets 12% for clothing. What amount does Pedro allow for his clothing budget?

Answer ____________________

20. Neil has a balance in his checking account of $639.47. If he writes a check for $249.44 and makes a deposit of $519.08, what will his new balance be?

Answer ____________________

21. Ingrid has $4500 in a 1-year CD at 6.4% interest. How much interest will she receive at the end of 1 year?

Answer ____________________

22. A farmer borrowed $4700 for 6 months at 12% interest. How much interest will the farmer pay on the loan?

Answer ____________________

Solve.

23. Rhonda borrowed $2300 to buy a stereo system. She will repay the loan in 24 installments of $109.37 each. How much interest will she pay on the loan?

Answer ______________________

24. Leland has an unpaid balance on his credit card of $450. If the finance rate is 1.8% per month, what is the finance charge on his balance?

Answer ______________________

25. Find the sticker price of a new car that has a base price of $9257, destination charge of $197, and these options:

Air conditioning	$797
Luggage rack	$149

Answer ______________________

26. Myron rented a compact car for $29 per day for 4 days and drove 540 miles. He was allowed 150 free miles per day, and was charged $0.20 for each additional mile. What was the total rental cost?

Answer ______________________

27. The yearly base premium for Heather's car insurance is $245.20. If her driver-rating factor is 2.4, how much is her yearly premium?

Answer ______________________

28. Suppose gasoline costs $1.19 per gallon. If your car gets 25 miles per gallon, find the cost of gasoline for a 700-mile trip.

Answer ______________________

29. A car was purchased for $10,900. Its value depreciated 45% in the first two years. Find the approximate trade-in value after two years.

Answer ______________________

30. Spencer can buy a monthly train pass for $76 or single-ride tickets for $2.50 each. He rides the train to and from work 20 days per month. How much money would he save each month by buying a monthly pass?

Answer ______________________

31. Greta buys a house for $70,000 and makes a down payment of $15,000. The mortgage payment is $11.01 per thousand of the loan amount. What is her monthly payment?

Answer ______________________

32. Last month Bridget's electric meter read 4205 kilowatt-hours. This month it reads 4840 kilowatt-hours. At 8¢ per kilowatt-hour, what will Bridget's electric bill be for the month?

Answer ______________________

Solve.

33. Bob wants to buy $70,000 worth of insurance for his new house. If the rate is $6.50 per $1000 of value, what will the insurance premium for his house be?

Answer ____________________

34. Estimate the cost of roofing a building if the roof area is 2100 square feet. The shingles will cost $68 per square and the labor charge is $75 per square. (One square of shingles covers 100 square feet.)

Answer ____________________

35. Torrey needed to hire a plumber to install a sink and a shower. The plumber charged for 3 hours and 15 minutes of labor. The rate was $45 for the first hour and $10 for each additional 15 minutes. How much did Torrey pay for the labor charge?

Answer ____________________

36. By paying cash instead of charging, Corwin bought a $650 desk for $611. What percent did he save?

Answer ____________________

37. Hannah's monthly salary is $1952. How much is withheld from Hannah's pay each month for social security tax? Use a social security tax rate of 7.65%.

Answer ____________________

38. Crystal is filing form 1040EZ. Her adjusted gross income was $20,508. She takes a $3000 standard deduction and a $2000 personal exemption. What is her taxable income? If the tax rate is 15%, what is Crystal's tax liability?

Answer ____________________

39. During the past year, $1964 of Chester's pay was withheld for federal income tax. If his tax liability is $2083, will he owe or get a refund? How much?

Answer ____________________

40. The market value of a home is $68,000 and the rate of assessment is 40%. The tax rate is 48 mills ($48 per $1000). What is the yearly real estate tax?

Answer ____________________

41. Xavier is self-employed and made a profit of $26,340 last year. He must pay $3651 in income tax. He also must pay social security tax of 13.3% of his profit. What is Xavier's total tax liability? Round to the nearest dollar.

Answer ____________________

42. Lana bought three items of clothing costing $15.35, $30.99, and $9.87. The sales tax rate was 8.5%. How much sales tax did she pay on the three items? What was the total cost?

Answer ____________________

Name ______________________ Date ______________

MASTERY TEST

Find each answer.

	a	*b*	*c*	*d*	*e*
1.	$2961 + 492$	$\$27.03 + 6.49$	$7 + 1.52$	$3872 + 390 + 6815$	$0.03 + 3.33 + 33.03$
2.	$27{,}406 - 8{,}528$	$\$63.90 - 34.05$	$25 - 2.731$	$60{,}000 - 975$	$7064 - 195$
3.	82×27	2.39×4	$\$5.21 \times 39$	7.4×0.6	$\$19.99 \times 23$
4.	$32\overline{)1920}$	$6\overline{)\$18.54}$	$8\overline{)7.24}$	$2.4\overline{)69.84}$	$0.95\overline{)4}$

Change each measurement.

	a	*b*	*c*
5.	17 m = ______ cm	72 kg = ______ g	3.9 L = ______ mL
6.	4 lb = ______ oz	41 in. = ______ ft ______ in.	24 pt = ______ c

Find each answer. Simplify.

	a	*b*	*c*	*d*
7.	$3 \times \frac{4}{9}$	$\frac{4}{5} \times \frac{3}{4}$	$2\frac{1}{6} \times \frac{2}{13}$	$\frac{2}{5} \times 7\frac{1}{2}$
8.	$\frac{1}{5} \div 10$	$\frac{5}{6} \div \frac{2}{3}$	$3\frac{1}{2} \div \frac{7}{12}$	$7\frac{1}{2} \div 1\frac{4}{5}$
9.	$\frac{3}{4} + \frac{1}{6}$	$2\frac{1}{3} + 5\frac{7}{12}$	$3 + 1\frac{7}{8}$	$4\frac{7}{10} + \frac{3}{5}$
10.	$\frac{5}{6} - \frac{2}{5}$	$5 - \frac{3}{7}$	$14 - 6\frac{2}{3}$	$7\frac{1}{5} - 3\frac{1}{4}$

Name ______________________ Date ______________

Find each number.

	a	*b*
11.	70% of 360	What percent of 1200 is 384?
12.	6% of what number is 18?	45% of 100

Solve.

13. Karl earns $256.00 for a 40-hour work week. He earns $9.60 for each hour of overtime. How much does he earn for a 45-hour week?

Answer ______________

14. Maria is a telephone technician who earns a weekly salary of $557. What is her yearly salary?

Answer ______________

15. Albert sells furniture. He earns a salary of $226 per week plus commission of 5% on all sales over $2400. One week he had sales of $9500. How much did he earn that week?

Answer ______________

16. In one pay period, Jackie had $440.20 in regular earnings. Her employer withheld $85.93 for taxes and $52.91 for insurance. What was Jackie's take-home pay for that pay period?

Answer ______________

17. Yvonne's disability insurance pays 65% of her salary for up to 26 weeks. If her weekly salary is $468 and she is disabled for 20 weeks, how much disability pay will she get?

Answer ______________

18. A store owner bought baskets for $2.45 each and sold them for $6.25. The overhead was $2.50 per basket. What was the profit on each basket?

Answer ______________

19. Mitch has a yearly income of $28,000. He budgets 10% for clothing. What amount does Mitch allow for his clothing budget?

Answer ______________

20. Brent has a balance in his checking account of $402.97. If he writes a check for $148.77 and makes a deposit of $245.11, what will his new balance be?

Answer ______________

21. Nicole has $8700 in a 1-year CD at 3.1% interest. How much interest will she receive at the end of 1 year?

Answer ______________

22. A farmer borrowed $7200 for 3 months at 10% interest. How much interest will the farmer pay on the loan?

Answer ______________

Name ______________________ Date ______________

Solve.

23. Ruth borrowed $1500 to buy a copy machine. She will repay the loan in 24 installments of $71.33 each. How much interest will she pay on the loan?

Answer ______________

24. Lloyd has an unpaid balance on his credit card of $700. If the finance rate is 1.9% per month, what is the finance charge on his balance?

Answer ______________

25. Find the sticker price of a new car that has a base price of $11,975, destination charge of $238, and these options:

Air conditioning	$650
Cruise control	$179

Answer ______________

26. Tom rented a full-size car for $39 per day for 4 days and drove 850 miles. He was allowed 150 free miles per day, and was charged $0.20 for each additional mile. What was the total rental cost?

Answer ______________

27. The yearly base premium for Lindsay's automobile insurance is $321.60. If her driver-rating factor is 2.6, how much is her yearly premium?

Answer ______________

28. Suppose gasoline costs $1.35 per gallon. If your car gets 30 miles per gallon, find the cost of gasoline for a 1200-mile trip.

Answer ______________

29. A car was purchased for $12,900. Its value depreciated 57% in the first three years. Find the approximate trade-in value after three years.

Answer ______________

30. Sammy can buy a monthly train pass for $42 or single-ride tickets for $1.50 each. He rides the train to and from work 18 days per month. How much money would he save each month by buying a monthly pass?

Answer ______________

31. Gina buys a house for $80,000 and makes a down payment of $18,000. The mortgage payment is $11.01 per thousand of the loan amount. What is her monthly payment?

Answer ______________

32. Last month Bonnie's electric meter read 3197 kilowatt-hours. This month it reads 3625 kilowatt-hours. At 7¢ per kilowatt-hour, what will Bonnie's electric bill be for the month?

Answer ______________

Name ______________________ Date ______________

Solve.

33. Ralph wants to buy $85,000 worth of insurance for his new house. If the rate is $7.20 per $1000 of value, what will the insurance premium for his house be?

Answer ______________

34. Estimate the cost of roofing a building if the roof area is 2400 square feet. The shingles will cost $59 per square and the labor charge is $65 per square. (One square of shingles covers 100 square feet.)

Answer ______________

35. Ernest needed to hire a plumber to install a water heater. The plumber charged for 1 hour and 30 minutes of labor. The rate was $54 for the first hour and $10 for each additional 15 minutes. How much did Ernest pay for the labor charge?

Answer ______________

36. By paying cash instead of charging, Irving bought a $380 cabinet for $361. What percent did he save?

Answer ______________

37. Nancy's weekly salary is $769. How much is withheld from Nancy's pay each week for social security tax? Use a social security tax rate of 7.65%.

Answer ______________

38. Robin is filing form 1040EZ. Her adjusted gross income was $18,534. She takes a $3000 standard deduction and a $2000 personal exemption. What is her taxable income? If the tax rate is 15%, what is Robin's tax liability?

Answer ______________

39. During the past year, $1287 of Oscar's pay was withheld for federal income tax. If his tax liability is $1054, will he owe or get a refund? How much?

Answer ______________

40. The market value of a home is $79,000 and the rate of assessment is 35%. The tax rate is 46 mills ($46 per $1000). What is the yearly real estate tax?

Answer ______________

41. Josh is self-employed and made a profit of $23,275 last year. He must pay $2351 in income tax. He also must pay social security tax of 13.3% of his profit. What is Josh's total tax liability? Round to the nearest dollar.

Answer ______________

42. Lilian bought three videotapes costing $12.85, $18.99, and $23.49. The sales tax rate was 8%. How much sales tax did she pay on the three videotapes? What was the total cost?

Answer ______________

WHERE TO GO FOR HELP

The table below lists the problems in the Mastery Test and the pages on which the corresponding skills and concepts are taught and practiced. If you missed one or more problems, find the page number or numbers that correspond to the number of each problem missed. Then turn to these pages, review the teaching, and practice the skills by doing one or more of the problems. Then correct the problems that you missed on this test.

PROBLEMS	PAGES	PROBLEMS	PAGES	PROBLEMS	PAGES	PROBLEMS	PAGES
1a	15	4e	23	10a	37	24	102
1b	16	5a	27	10b	38	25	104
1c	16	5b	27	10c	38	26	107
1d	15	5c	27	10d	38	27	111
1e	16	6a	34	11a	48–49	28	112
2a	15	6b	34	11b	51	29	117
2b	16	6c	34	12a	50	30	120
2c	16	7a	42	12b	48–49	31	128
2d	15	7b	42	13	59	32	131
2e	15	7c	42	14	60	33	132
3a	18	7d	42	15	63	34	140
3b	19	8a	43	16	66	35	141
3c	19	8b	43	17	76	36	146
3d	19	8c	43	18	71	37	150
3e	19	8d	43	19	85	38	152–154
4a	22	9a	35	20	87	39	152–156
4b	22	9b	36	21	90	40	165
4c	23	9c	36	22	97	41	180
4d	23	9d	36	23	101	42	170

ANSWER KEY

PRETEST

Page 6

	a	*b*	*c*	*d*	*e*
1.	30	90	570	3250	12,590
2.	8.4	25.8	6.2	53.2	24.1

Answers may vary when estimates are required.

	a	*b*	*c*	*d*
3.	36,500	4800	28	26
4.	8	9	10	0.7

	a	*b*	*c*	*d*	*e*
5.	9716	$52.62	17.36	7580	60.65
6.	7843	$83.01	23.583	33,101	7878
7.	1620	22.59	$232.68	2.28	$2626.14
8.	52 R20	$7.09	0.6	32.1	4.8

Page 7

	a	*b*	*c*	*d*
9.	$\frac{4}{7}$	$\frac{5}{7}$	$\frac{2}{3}$	$\frac{3}{4}$
10.	$\frac{4}{6}, \frac{5}{6}$	$\frac{9}{12}, \frac{4}{12}$	$\frac{15}{24}, \frac{20}{24}$	$\frac{15}{18}, \frac{14}{18}$

	a	*b*	*c*
11.	2300	46,000	5400
12.	80	2 ft 5 in.	60

	a	*b*	*c*	*d*
13.	$\frac{17}{18}$	$9\frac{2}{3}$	$10\frac{5}{8}$	$7\frac{1}{4}$
14.	$\frac{5}{12}$	$6\frac{3}{8}$	$6\frac{3}{5}$	$3\frac{11}{12}$
15.	$2\frac{1}{2}$	$\frac{2}{3}$	2	3
16.	$\frac{2}{27}$	$3\frac{1}{2}$	$2\frac{1}{2}$	$1\frac{47}{51}$

Page 8

	a	*b*
17.	200	45%
18.	80	130

19. $403.75 **20.** $30,160
21. $596 **22.** $463.96
23. $7920 **24.** $51.00
25. $22,400 **26.** $568.96
27. $134.40 **28.** $168

Page 9

29. $105.60 **30.** $12
31. $16,248 **32.** $237
33. $228.84 **34.** $35.40
35. $500.00 **36.** $41.60
37. $594.54 **38.** $30.88

Page 10

39. $575 **40.** $3400
41. $62 **42.** 20%
43. $68.24 **44.** $10,412; $1561.80
45. Owe $21.00 **46.** $1444
47. $8008 **48.** $8.10; $143.07

UNIT 1

Page 12

	millions	hundred thousands	ten thousands	thousands	hundreds	tens	ones	tenths	hundredths	thousandths
1.	7,	6	4	6,	9	1	8.	7	6	4
2.		1	8	9,	4	9	8.	0	0	4
3.			9	6,	9	5	0.	5	3	
4.				2,	0	0	0.	9	9	

5. 99.01
6. 56,047.31
7. $42.53
8. 8.612
9. fifty and eight tenths
10. eight hundred sixteen and seven hundredths
11. three hundred eighty-five and forty-one hundredths
12. one thousand five and seventy-five thousandths
13. thirteen dollars and ninety-nine cents
14. two hundred ten dollars and thirty-five cents

Page 13

	a	b	c
1.	>	=	>
2.	>	<	=
3.	>	<	>
4.	<	<	<

5. 45 76 98
6. 287 789 891
7. 190 561 754
8. 1256 3963 9867
9. 2876 8432 8765
10. 32.85 78.67 84.13
11. 24.678 375.2 973.9
12. 0.340 0.674 0.970
13. 12.97 12.975 12.976
14. 4.5 7.6 9.8
15. 0.190 1.90 190

Page 14

	a	b	c	d
1.	50	90	90	20
2.	480	310	3260	4190
3.	500	700	400	800
4.	4400	5400	11,800	68,900
5.	9.3	7.3	3.6	45.7
6.	1.1	4.9	22.8	39.1
7.	2.83	15.95	6.30	81.61
8.	4.90	0.61	1.83	0.10
9.	$24	$25	$40	$96
10.	$330	$836	$1121	$2000

Page 15

	a	b	c	d	e
1.	706	381	664	807	1344
2.	9794	105,747	103,538	6901	48,266
3.	17,549	178,404	73,556	18,705	104,025
4.	608	514	7508	1457	1302
5.	10,971	47,777	51,497	77	9525
6.	67,225	47,196	83,919	43,178	5001

	a	b	c
7.	6607	399	594,434

Page 16

	a	b	c	d	e
1.	12.67	98.5	108.6	9.7	8.6
2.	$3.97	$893.83	$79.94	$29.95	$2.56
3.	2038.23	21,059.67	3802.653	115.910	73.03
4.	1.4	3.77	4.9	6.06	6.9
5.	$891.32	$2260.46	$550.65	$175.77	$262.42
6.	9023.54	3.643	6.227	52.643	1.903

	a	b	c
7.	65.39	689.3	1517.67

Page 17

	a	b	c	d
1.	190	10,670	4440	83,520
2.	8500	81,400	2100	2700
3.	$125.00	$624.00	$25.00	$19.00
4.	89.3	124.7	2.3	2.5
5.	2.41	9.22	242.68	42.74

Page 18

	a	b	c	d	e
1.	72	84	3208	47,684	809,280
2.	520	2436	2124	45,948	280,000
3.	14,508	9200	36,652	12,510	26,037
4.	334,900	439,008	104,272	70,200	177,786

	a	b	c
5.	59,157	26,676	5152
6.	20,500	2850	1920

Page 19

	a	b	c	d	e
1.	85.25	5.36	88.78	65,241.75	2342.22
2.	3.750	$55.84	$3605.06	9720.00	18.13
3.	0.12006	0.000882	0.003258	2.773278	0.470764

	a	b	c
4.	6.8742	17.355	5807.43
5.	5.2026	532.684	11,669.2

Page 20

1. To solve, subtract.
   $483 − $37 = $446
2. Multiply
   $21,482

Page 21

3. Multiply
   234 miles
4. Divide
   54 pairs
5. Multiply
   $1944
6. Subtract
   564 more miles
7. Divide
   $419
8. Add
   116.5 pounds
9. Subtract
   $47.88
10. Multiply and add $48.64
11. Divide
    $7.54
12. Subtract
    41.5 kilograms

Page 22

1.

```
      a                b
     12 R27           23 R14
45 ) 567        23 ) 543
   - 45↓           - 46↓
     117              83
   -  90            - 69
      27              14
```

c	*d*
7 R30	540 R9

	a	*b*	*c*	*d*
2.	14 R1	25 R3	7 R32	260 R27
3.	122 R57	55 R5	219 R5	72
4.	402	101	60 R3	201 R23

	a	*b*	*c*
5.	448 R1	27 R11	74 R7
6.	105 R8	83 R44	141 R6

Page 23

	a	*b*	*c*	*d*
1.	0.8	7.1	1.0	5.4
2.	5.8	30	2.8	9.7
3.	83	1.3	5912	218
4.	0.0	0.1	19.4	0.1

	a	*b*	*c*
5.	0.13	7.3	187
6.	29.8	0.1	219

Page 24

b

Answers may vary.

```
1.  361  →   360
   × 55  →  × 60
            21,600
```

	a	*b*	*c*	*d*
1.	42,500	21,600	400	82,800
2.	5.4	120	8.0	75.2
3.	120	120	100	230
4.	0.60	100	4	0.40
5.	15	4	30	9
6.	40	6	4	0.6

Page 25

1. 4.07	**2.** 407	**3.** 0.407
4. 284	**5.** 28.4	**6.** 0.284
7. 3.912	**8.** 39.12	**9.** 84
10. 8.4	**11.** 34,010	**12.** 0.45

13. 2000 centimeters
14. 43,200 millimeters
15. 3700 milligrams
16. 0.734 kilograms

Page 26

1. To solve, multiply.

```
   59
 × 14
  236
  59
  826 feet
```

The total length of the train without the engine is 826 feet.

2. 135,850 kilograms
3. $309.60
4. 5.1 feet
5. $17,447.00
6. 65 bushels
7. $111.44
8. 420 miles per hour

Page 27

	a	*b*	*c*
1.	5000	1270	97,000
2.	460	12,000	8000
3.	48	98	2.1
4.	50	17	3

Page 28

1. David worked 8.5 hours on Monday, 6 hours on Tuesday, ~~and 7.5 hours on Wednesday.~~ How much longer did he work on Monday than Tuesday? 2.5 hours
2. Winona bought a car that cost $9880, ~~including a $199 service warranty.~~ She paid $1000 as a down payment and took out a loan for the rest. What was the amount of her loan? $8880
3. Raul works part-time ~~and earns $75 per week.~~ If he saves about $25 per week, how much money will he save in 6 weeks? $150
4. Kris spent $4.32 for a calculator, ~~$12.89 for a backpack,~~ and $8.99 for a lantern. How much more did the lantern cost than the calculator? $4.67

Page 29

5. ~~A ream of paper contains 500 sheets.~~ If a person buys 12 reams for $5.25 each, what is the total cost? $63
6. A ~~16-inch pizza~~ costs $13.95. If five friends share the cost equally, how much does each friend pay? $2.79
7. ~~The original price of a jogging suit was $38.50.~~ Tommy bought the suit on sale for $28.95. He also spent $6.29 for gloves. How much did Tommy spend for the jogging suit and gloves? $35.24

8. Martha is 63 inches tall ~~and weighs 115 pounds.~~ Her sister is 67 inches tall ~~and weighs 122 pounds.~~ How much shorter is Martha than her sister?
4 inches
9. Jaleesa ~~paid $2.69~~ for an 8-foot strip of molding. If she cut off 2 feet 4 inches of molding, find the length of the remaining piece. 5 feet 8 inches
10. The distance from Greg's house to school is 4.2 kilometers. ~~The distance to the library is 2.7 kilometers.~~ If Greg rides his bike from home to school and back, how far does he ride?
8.4 kilometers
11. Evelyn worked 4 hours 15 minutes before lunch, ~~then took a break for 45 minutes.~~ She worked 3 hours 30 minutes after lunch. How much time was she working that day? 7 hours 45 minutes
12. Ms. Benitez works part-time and earns $44 per week. ~~She earns $5.50 per hour.~~ How much does she earn in 12 weeks? $528
13. Brad drove ~~134 miles in~~ 2 hours 20 minutes. How many minutes did he drive? 140 minutes
14. A carton of milk contains 250 milliliters ~~and costs 39¢.~~ How many cartons could be filled from 4 liters of milk? 16 cartons

Page 30

1. To solve, add.

$$\begin{array}{r} \$4.48 \\ 1.88 \\ 0.87 \\ +\ 3.89 \\ \hline \$11.12 \end{array}$$

The total for all of the items is $11.12.

2. $84.64
3. $5.49
4. $14.08
5. $55.47
6. $2.49 per pound
7. 80 km per hr
8. $0.23

Page 31 Unit 1 Review

	a	*b*	*c*	*d*
1.	366	110.1	52,668	388.82
2.	78.53	0.4	1985	291.9
3.	37,770	15,004	0.567	862.69
4.	1825	2.3733	193,404	395.901
5.	1122	0.115	69,300	1044

Answers may vary.

	a	*b*	*c*
6.	620	16	900
7.	2400	4	1200

	a	*b*	*c*
8.	3700	397,000	1900

UNIT 2

Page 32

	a	*b*	*c*
1.	$\frac{2}{5}$	$\frac{1}{9}$	$\frac{5}{60}$
2.	two thirds	seven tenths	one eighty-eighth
3.	$\frac{7}{12}$; seven twelfths	$\frac{5}{12}$; five twelfths	$\frac{11}{12}$; eleven twelfths

	a	*b*	*c*	*d*
4.	$\frac{2}{3}$	$\frac{1}{2}$	$\frac{2}{3}$	$\frac{4}{5}$
5.	1	$\frac{1}{2}$	$\frac{2}{3}$	$\frac{6}{7}$
6.	$\frac{15}{20}$; $\frac{4}{20}$	$\frac{3}{6}$; $\frac{4}{6}$	$\frac{21}{30}$; $\frac{4}{30}$	$\frac{22}{24}$; $\frac{15}{24}$

Page 33

	a	*b*	*c*
1.	1	1	1
2.	3	4	6
3.	$7\frac{2}{5}$	$2\frac{2}{5}$	$7\frac{5}{7}$
4.	$7\frac{3}{8}$	$3\frac{7}{8}$	$6\frac{3}{4}$
5.	$8\frac{1}{9}$	$9\frac{1}{16}$	$11\frac{8}{9}$

	a	*b*	*c*
6.	$\frac{3}{1}$	$\frac{7}{1}$	$\frac{2}{1}$
7.	$\frac{28}{5}$	$\frac{30}{7}$	$\frac{160}{11}$
8.	$\frac{90}{11}$	$\frac{174}{7}$	$\frac{43}{10}$
9.	$\frac{16}{9}$	$\frac{22}{7}$	$\frac{76}{5}$
10.	$\frac{56}{9}$	$\frac{38}{7}$	$\frac{14}{9}$

Page 34

1. *a* 98 in. = 8 ft 2 in.
1 foot = 12 inches

$$\begin{array}{r} 8\ \text{R}2 \\ 12\,\overline{)\,98} \\ -\ 96 \\ \hline 2 \end{array}$$

b 37 in. = 3 ft 1 in.
1 foot = 12 inches

$$\begin{array}{r} 3\ \text{R}1 \\ 12\,\overline{)\,37} \\ -\ 36 \\ \hline 1 \end{array}$$

	a	*b*
2.	1 lb 12 oz	4 lb 0 oz
3.	4 qt 1 pt	7 gal 3 qt

	a	*b*	*c*
4.	72	180	10,560
5.	48	320	28,000
6.	33	20	32

Page 35

	a	*b*	*c*	*d*
1.	1	$1\frac{1}{13}$	1	$\frac{13}{17}$
2.	$\frac{9}{11}$	$\frac{11}{15}$	$\frac{3}{4}$	$\frac{94}{99}$
3.	$1\frac{5}{14}$	$1\frac{4}{15}$	$1\frac{1}{2}$	$1\frac{1}{6}$
4.	$\frac{23}{24}$	$\frac{7}{12}$	$1\frac{7}{12}$	$1\frac{4}{15}$

	a	*b*	*c*
5.	$\frac{13}{21}$	$1\frac{7}{15}$	$1\frac{19}{36}$
6.	$1\frac{3}{10}$	$1\frac{1}{12}$	$\frac{44}{45}$

Page 36

	a	*b*	*c*	*d*
1.	$7\frac{11}{12}$	$7\frac{1}{4}$	$11\frac{1}{3}$	$5\frac{3}{5}$
2.	$11\frac{2}{3}$	$10\frac{1}{10}$	$39\frac{5}{14}$	$35\frac{1}{8}$
3.	$14\frac{5}{9}$	$47\frac{1}{5}$	$30\frac{1}{4}$	$50\frac{5}{8}$
4.	$6\frac{11}{20}$	$49\frac{13}{18}$	$6\frac{13}{21}$	$30\frac{1}{12}$
5.	$12\frac{11}{15}$	$10\frac{37}{45}$	$12\frac{1}{12}$	$93\frac{9}{40}$

	a	*b*	*c*
6.	$9\frac{39}{40}$	$13\frac{17}{99}$	$31\frac{1}{15}$

Page 37

	a	*b*	*c*	*d*
1.	$\frac{1}{3}$	$\frac{2}{9}$	$\frac{1}{5}$	$\frac{1}{4}$
2.	$\frac{1}{4}$	$\frac{11}{16}$	$\frac{9}{14}$	$\frac{1}{8}$
3.	$\frac{3}{8}$	$\frac{8}{15}$	$\frac{5}{9}$	$\frac{7}{9}$
4.	$\frac{8}{21}$	$\frac{26}{45}$	$\frac{7}{12}$	$\frac{11}{24}$
5.	$\frac{2}{45}$	$\frac{5}{18}$	$\frac{3}{56}$	$\frac{3}{14}$

	a	*b*	*c*
6.	$\frac{5}{33}$	$\frac{1}{10}$	$\frac{5}{36}$

Page 38

	a	*b*	*c*	*d*
1.	$\frac{5}{12}$	$5\frac{3}{8}$	$21\frac{7}{9}$	$7\frac{8}{11}$
2.	$1\frac{1}{2}$	$2\frac{2}{3}$	$2\frac{1}{5}$	$2\frac{5}{7}$
3.	$7\frac{1}{6}$	$1\frac{1}{6}$	$12\frac{1}{8}$	$3\frac{3}{4}$
4.	$6\frac{1}{21}$	$13\frac{8}{15}$	$5\frac{1}{6}$	$4\frac{11}{20}$
5.	$6\frac{19}{24}$	$3\frac{7}{18}$	$2\frac{11}{15}$	$2\frac{13}{20}$

	a	*b*	*c*
6.	$5\frac{17}{18}$	5	$7\frac{3}{5}$

Page 39

1. To solve, add.

$$1\frac{1}{2} = 1\frac{2}{4}$$
$$+\ 1\frac{3}{4} = 1\frac{3}{4}$$
$$2\frac{5}{4} = 2 + 1\frac{1}{4} = 3\frac{1}{4} \text{ dozen eggs}$$

They used $3\frac{1}{4}$ dozen eggs in all.

2. $\frac{1}{2}$ the trip

3. $1\frac{5}{12}$ cups

4. $9\frac{3}{8}$ pounds

5. $\frac{7}{12}$ foot

6. $10\frac{1}{5}$ pounds

7. $6\frac{1}{8}$ hours

8. Raj; $\frac{1}{8}$ mile

Page 40

1.

Number of Color Pencils (columns); Number of Erasers (rows)

Erasers \ Pencils	0	1	2	3	4	5
0						
1		$1.10	$1.50	$1.90	$2.30	$2.70
2		$1.80	$2.20	$2.60	$3.00	$3.40
3		$2.50	$2.90	$3.30	$3.70	$4.10
4		$3.20	$3.60	$4.00	$4.40	$4.80
5		$3.90	$4.30	$4.70	$5.10	$5.50

2. 3 **3.** 5

Page 41

4.

Plain Yogurt (columns); Fruit Yogurt (rows)

Fruit Yogurt \ Plain Yogurt	0	1	2	3
0	0	$0.99	1.98	2.97
1	$1.19	2.18	3.17	4.16
2	2.38	3.37	4.36	5.35
3	3.57	4.56	5.55	6.54

5. 3 plain, 3 fruit **6.** 2 fruit, 1 plain

7.

Dimes	Pennies	Total Value	
22	3	$2.23	No
17	8	$1.78	No
12	13	$1.33	Yes

13 pennies

8.

Quarters	Dimes	Total Value	
1	9	$1.15	No
2	8	$1.30	No
3	7	$1.45	No
4	6	$1.60	Yes

4 quarters, 6 dimes

9.

Folders 50¢	Pencils 20¢	Rulers 10¢	Total
2	0	0	$1.00
1	2	1	$1.00
1	1	3	$1.00
1	0	5	$1.00
0	5	0	$1.00
0	4	2	$1.00
0	3	4	$1.00
0	2	6	$1.00
0	1	8	$1.00
0	0	10	$1.00

Page 42

	a	*b*	*c*	*d*
1.	$\frac{6}{25}$	$\frac{35}{81}$	$\frac{21}{64}$	$\frac{18}{49}$
2.	$\frac{8}{21}$	$\frac{7}{20}$	$\frac{4}{7}$	$\frac{18}{35}$
3.	$3\frac{1}{5}$	$\frac{2}{3}$	$4\frac{2}{7}$	$1\frac{1}{2}$
4.	$3\frac{3}{7}$	$2\frac{43}{45}$	$51\frac{3}{4}$	57
5.	40	$55\frac{13}{35}$	$24\frac{1}{9}$	$40\frac{25}{32}$

	a	*b*	*c*
6.	$\frac{4}{9}$	$\frac{1}{18}$	$\frac{1}{35}$

Page 43

	a	*b*	*c*	*d*
1.	$\frac{2}{3}$	1	2	5
2.	$\frac{27}{35}$	$6\frac{2}{3}$	$\frac{6}{7}$	$\frac{1}{2}$
3.	$2\frac{2}{5}$	$\frac{8}{45}$	5	$\frac{1}{8}$
4.	$1\frac{15}{29}$	$1\frac{11}{12}$	$\frac{21}{25}$	$\frac{11}{24}$
5.	$9\frac{3}{8}$	$\frac{1}{18}$	$6\frac{3}{4}$	$\frac{2}{45}$
6.	$1\frac{95}{153}$	$1\frac{55}{141}$	$\frac{2}{3}$	$\frac{3}{10}$

Page 44

	a	*b*
1.	$32\% = 0.32$ $32\% = \frac{32}{100} = \frac{8}{25}$	0.05; $\frac{1}{20}$

	a	*b*		*a*	*b*
2.	0.04; $\frac{1}{25}$	0.95; $\frac{19}{20}$	**6.**	0.15; $\frac{3}{20}$	0.03; $\frac{3}{100}$
3.	0.21; $\frac{21}{100}$	0.60; $\frac{3}{5}$	**7.**	0.89; $\frac{89}{100}$	0.75; $\frac{3}{4}$
4.	0.01; $\frac{1}{100}$	3.00; $\frac{300}{100}$	**8.**	0.90; $\frac{9}{10}$	0.42; $\frac{21}{50}$
5.	1.00; $\frac{100}{100}$	0.07; $\frac{7}{100}$			

Page 45

	a	*b*	*c*
1.	19%	30%	65%
2.	1%	10%	90%
3.	75%	18%	9%
4.	5%	18.3%	20%
5.	10%	60%	41%
6.	20%	90%	50%
7.	70%	75%	25%
8.	$6\frac{1}{4}\%$	$37\frac{1}{2}\%$	$12\frac{1}{2}\%$

Page 46

1. 0.10; 10%
2. 0.50; 50%
3. $\frac{3}{4}$; 0.75; 75%
4. $\frac{3}{10}$; 0.30; 30%
5. $\frac{9}{20}$; 0.45; 45%
6. $\frac{1}{5}$; 0.20; 20%
7. 0.265
8. $\frac{1}{5}$
9. 90%
10. 0.08
11. $\frac{1}{2}$
12. 0.40

Page 47

	a	*b*	*c*
1.	6	2.505	2.75
2.	2.25	1.25	1.875
3.	1.01	3	1.99
4.	0.0025	0.004	0.001
5.	0.003125	0.00625	0.008
6.	0.00125	0.0075	0.003

Page 48

	a	*b*
1.	9	20
2.	33.6	12
3.	70	9
4.	6.4	70

5. $5.67 **6.** $4.68

Page 49

a *b*

1. 50% of 60 — $\frac{1}{4}$, 100

$50\% = \frac{50}{100} = \frac{1}{2}$

$\frac{1}{2} \times 60 = 30$

	a	*b*
2.	$\frac{3}{4}$, 90	$\frac{1}{3}$, 15
3.	$\frac{1}{10}$, 18	$\frac{1}{5}$, 12
4.	$\frac{2}{5}$, 60	$\frac{3}{5}$, 96

5. $17.04 6. $3200

Page 50

	a	*b*		*a*	*b*
1.	40	250	4.	80	2000
2.	17	5	5.	15	5
3.	100	300	6.	160	15

Page 51

	a	*b*		*a*	*b*
1.	25%	50%	4.	1%	50%
2.	2%	9%	5.	10%	10%
3.	$33\frac{1}{3}\%$	20%	6.	50%	1%

Page 52

1. original cost → $85
percent of decrease → × 0.20
amount of decrease → $17.00

original cost → $85
amount of decrease → − 17
sale price → $68

2. 495 people 3. $59.84 4. 36 people
5. 12 home runs 6. $560

Page 53

1. original value → $80,000
percent of increase → × 0.10
amount of increase → $8,000

original value → $80,000
amount of increase → 8,000
new value → $88,000 this year

2. $62.04 3. 29 people 4. $598
5. 57 words 6. 77 beats per minute

Page 55

1. $n + 0.73 = 9.38$
or $n = 9.38 - 0.73$
$8.65

2. $0.2 \times n = 10$
or $n = 10 \div 0.2$
50 students

3. $3 \times \frac{3}{4} = n$
$2\frac{1}{4}$ cups

4. $\frac{2}{3} \times n = 24$
or $n = 24 \div \frac{2}{3}$
36 hours

5. $1\frac{5}{8} + n = 2\frac{1}{4}$
or $n = 2\frac{1}{4} - 1\frac{5}{8}$
$\frac{5}{8}$ miles

6. $n \div 5 = 4\frac{1}{2}$
or $n = 4\frac{1}{2} \times 5$
$22\frac{1}{2}$ inches

7. $\frac{1}{10} \times n = 3$
or $n = 3 \div \frac{1}{10}$
30 students

8. $1.5 \times n = 24$
or $n = 24 \div 1.5$
16 inches

Page 56

1. 79% 2. 35% 3. 5% 4. 22%
5. 18% 6. 95%
7. 6 students 8. 12 students

Page 57 Unit 2 Review

	a	*b*	*c*
1.	96 oz	4 ft 4 in.	30 c
2.	15 c	1 lb 3 oz	15 ft

	a	*b*	*c*	*d*
3.	$\frac{1}{10}$	$\frac{3}{10}$	$1\frac{4}{9}$	$2\frac{2}{3}$
4.	$\frac{100}{161}$	$2\frac{1}{8}$	$\frac{5}{8}$	$10\frac{1}{9}$
5.	$1\frac{5}{8}$	$16\frac{25}{42}$	$5\frac{7}{8}$	$1\frac{1}{3}$

	a	*b*
6.	7.68	52.5
7.	2.5%	30%
8.	134.4	1300
9.	500	41.16

UNIT 3

Page 58

	Hours Worked	Weekly Earnings
1.	40	$268.80
2.	38	$322.24
3.	37.5	$195.00
4.	34.5	$252.89

5. Hours per day 8 × Days per week 5 = Hours per week 40

Total hours 40 × Hourly wage $12.50 = Pay per week $500
James earns $500 in one week.

6. $491.25 7. $8.04 8. $50.40

Page 59

1. $8.20 × 1.5 = $12.30
The overtime rate is $12.30.

2. $8.79 3. $18.99 4. $11.48

	Regular Earnings	Overtime Rate	Overtime Earnings	Total Earnings
5.	$412.00	$15.45	$92.70	$504.70
6.	$220.00	$8.25	$37.13	$257.13
7.	$595.20	$22.32	$223.20	$818.40
8.	$358.00	$13.43	$33.58	$391.58

9. $760.24

Page 60

	Weekly Salary	Yearly Salary	Monthly Salary
1.	$653.95	$34,005.40	$2833.78
2.	$738.85	$38,420.20	$3201.68
3.	$454.46	$23,631.92	$1969.33
4.	$484.62	$25,200.00	$2100.00
5.	$486.80	$25,313.60	$2109.47

6. Linda: $2538 per month × 12 months = $30,456 per year
Billie: $32,700 per year
Billie earns more.

7. $128.10 8. $33,550.00 9. $4160
10. Ted 11. $286.85

Page 61

1. $6.80 per hour × 40 hours = $272
$272
+ 240 tips
$512 total weekly earnings
The hair stylist's total weekly earnings are $512.

2. $420.00 3. $92.00 4. $255.00
5. $422.00 6. $120.50

Page 62

	Decimal	Amount of Commission
1.	0.03	$3750.00
2.	0.02	$1928.00
3.	0.025	$2500.00
4.	0.035	$2411.50
5.	0.015	$2611.50
6.	0.05	$1624.55

7. 25% = 0.25
Commission rate 0.25 × Sales $236 = Amount of commission $59
Serene earns $59 commission.

8. $119.93 9. $42.75 10. $393.20

Page 63

1. Commission: 0.02 × 149,000 = $2980
Salary plus commission: $800 + $2980 = $3780
Rita's total earnings in December were $3780.

2. $619.95 3. $604 4. $6305
5. Job B 6. About $400; $21,000

Page 64

1.

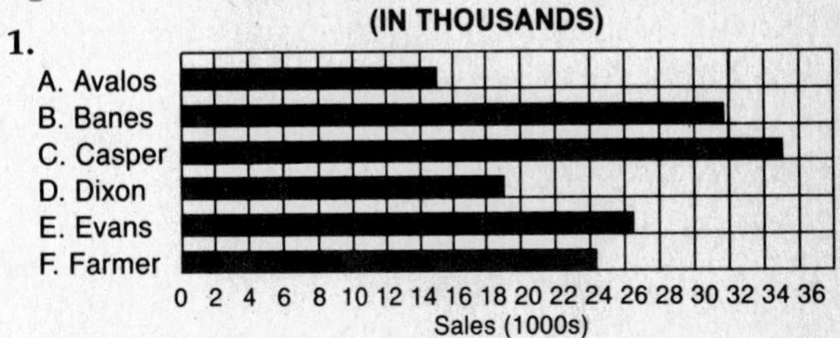

2. C. Casper 5. $2295
3. $5280 6. 4
4. A. Avalos

Page 65

	Monthly Premium	Employer's Contribution		Employee's Contribution	
		Monthly	Yearly	Monthly	Yearly
1.	$205	$102.50	$1230	$102.50	$1230
2.	$125	$112.50	$1350	$12.50	$150
3.	$160	$120.00	$1440	$40.00	$480
4.	$238	$47.60	$571.20	$190.40	$2284.80

5. Company A
Salary: $19,000
Medical coverage: $150 × 12 = $1800
Total: $20,800
Company B
Salary: $1700 × 12 = $20,400
Company A seems to be better.

6. $46.80

Page 66

	Total Ded.	Net Pay
1.	$153.30	$396.70
2.	$76.04	$273.96
3.	$102.98	$247.02
4.	$36.65	$172.55
5.	$268.94	$510.59
6.	$167.97	$490.47

7. $278.54

Page 67

	Nearest cent	Nearest dime	Nearest dollar
1.	$1.09	$1.10	$1.00
2.	$14.26	$14.30	$14.00
3.	$4.89	$4.90	$5.00
4.	$0.81	$0.80	$1.00
5.	$2.68	$2.70	$3.00
6.	$3.67	$3.70	$4.00
7.	$0.40	$0.40	$0.00
8.	$15.56	$15.60	$16.00

9. $8.30 **10.** $16.00

Page 68

1. Substep 1: Find amount of state tax deducted.
   Substep 2: Find total deductions.
2. Substep 1: Find earnings for the first week.
   Substep 2: Find earnings for the second week.

Page 69

3. $9.12
$65.36
$238.64

4. $241.20
$274.00
$32.80

5. $0.45 **6.** Evelyn; $680 more
7. $7.54 **8.** 8%
9. $119 **10.** 6 hours

Page 70

1. Gross income − Expenses = Profit
   $156,000 − ($97,000 + $26,000) = $33,000

2. $8400 **3.** $5000 **4.** 9.6%
5. $12,000 **6.** 21%

Page 71

1. Markup = 60% of $40 = $24
   Selling Cost = $40 + $24 = $64

2. $11.60

3. 64¢ per pound **4.** $300
5. $10 **6.** $10

Page 72

1.–2.

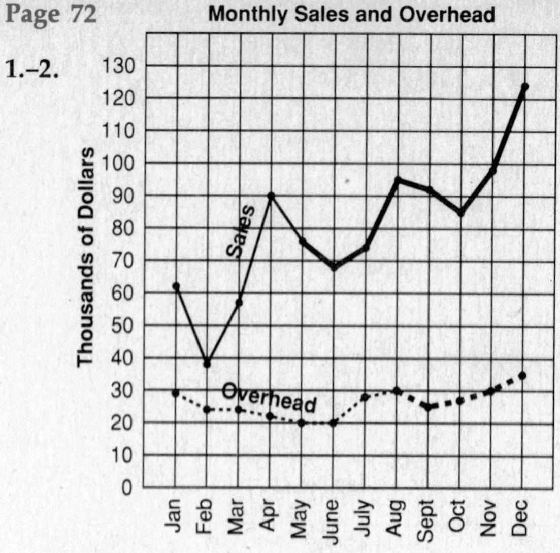

3. December **7.** Winter
4. February **8.** Fall
5. December **9.** March–April
6. May–June **10.** January–February

Page 73

	Total Expenses
1.	$10,474
2.	$1943
3.	$4082
4.	$8452
5.	$943
6.	$60

7.	8.	9.	10.	11.
$1943	$4082	$8452	$943	$60
$7757	$4368	$23,987	$3412	$2690

Page 74

	Points Each	Multiply Number of Employees by Points Each
1.	1	41
2.	2	50
3.	3	54
4.		TOTAL POINTS: 145

5. Divide $246,300 by 145 points.

```
       1 698.6
145 ) 246,300.0
     - 145
       1013
     - 870
       1430
     - 1305
        1250
      - 1160
          900
          870
```

To the nearest dollar, each point is worth $1699.

6. $3398 7. $5097 8. $5065.50
9. 58 points 10. $1497.88

Page 75

1. Number of nights × Rate per night
   4 × $30.75 = $123.00
   Hotel tax: 0.11 × $123.00 = $13.53
   Total cost including tax:
   $123.00 + $13.53 = $136.53

2. $76.50

	Breakfast	Lunch	Dinner	Totals
3.	$4.75	$5.50	$9.50	$19.75
4.	$3.85	$6.00	$11.50	$21.35
5.	$4.50	$4.95	$10.50	$19.95
6.	$13.10	$16.45	$31.50	$61.05

	Air Fares	Rental Cars & Gasoline	Totals
7.	$315	$30	$345
8.		$30	$30
9.	$275	$30 + $5 gas	$310
10.	$590	$95	$685

Page 76

1. Disability pay each week:
   50% × $380 = 0.5 × $380 = $190
   Total disability pay (25 weeks): 25 × $190 = $4750

2. $270 $9720
3. $760 $31,920 4. $289 $23,409
5. $471 $30,615 6. $984 $204,672

Page 77

1. Monthly income × Months per yr = Yearly income
   $460 × 12 = $5520

2. $17,508 3. $895.17 4. $4640
5. $573.60 6. $245

Page 78

1. $260 2. $9.49
3. $28,600 4. $37,440

Page 79

1. $2600 2. 14.3%
3. 12.5% 4. $330
5. $396 6. $4700

Page 80

1. 50%
2. 20%
3. 26% 4. 10% 5. 20%

6.

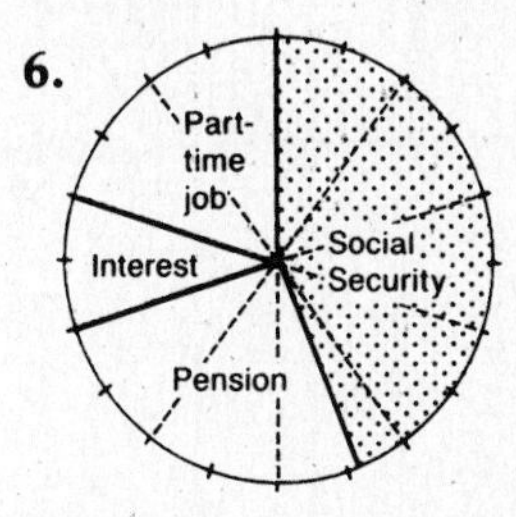

7. less
8. pension
9. part-time job

Page 81 Unit 3 Review

1. $881.92 2. $680.77
3. $345 4. $390.85
5. $6300 6. $1440

7.

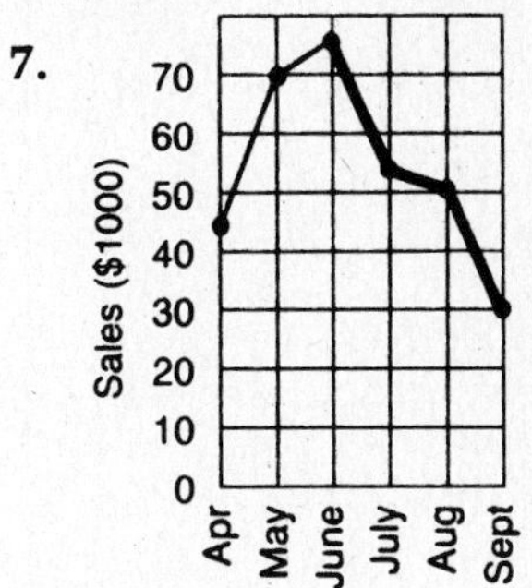

8. 67%; 33%

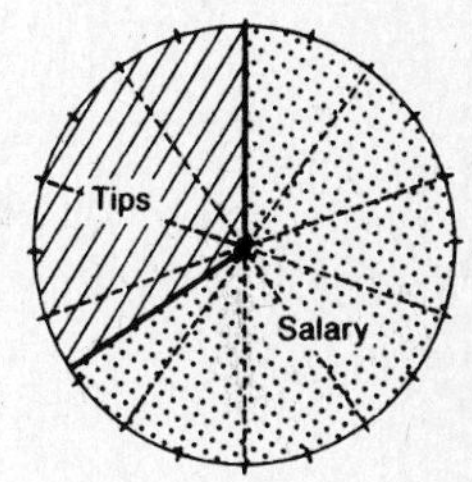

9. April and May 10. $3.00

UNIT 4

Page 82

	Amount of Discount	Discount Price
1.	$3.25	$29.25
2.	$5.93	$33.57
3.	$2.50	$7.50
4.	$4.40	$17.60
5.	$4.40	$4.40
6.	$4.16	$8.33

7. $7.00; less 8. $3.80; less
9. $20.00; more 10. $31.50; less

Page 83

	Date	Description	Receipts	Expenses: Food & Clothes	Housing & Utilities	Transpor-tation	Savings & Giving	Misc.	Balance $240.90
1.									
2.	1/10	Gasoline				14.50			$657.53
3.	1/12	Utility bills			85.79				571.74
4.	1/15	Car payment				257.00			314.74
5.	1/16	Groceries		78.11					236.63
6.	1/19	Wages	598.00						834.63
7.	1/22	Contributions					40.00		794.63
8.	1/24	Game tickets						16.40	778.23
9.	1/27	Savings					50.00		728.23
10.	1/30	Rent			540.00				188.23
11.		TOTALS	$1168.00	$179.98	$625.79	$271.50	$127.00	$16.40	

12. $1220.67; more

Page 84

1. $220 + $430 + $90 + $100 + $120 = $960 per month
   The total of Natalie's monthly expenses is $960.
2. Less
3. $110   4. No
5. $400   6. $120
7. $35   8. $133.33

Page 85

1. $674.70, $449.80, $155.70, $224.90, $224.90, $1730.00
2. $729.60, $661.20, $228.00, $342.00, $319.20, $2280.00
3. Moore
4. Moore
5. Pasko, $86.50; Moore, $114.00
6. Pasko, 6%; Moore, 4%

Page 86

1. Fifteen and $\frac{20}{100}$
2. Forty and $\frac{10}{100}$
3. Two hundredy forty-three and $\frac{08}{100}$
4. One hundred six and $\frac{90}{100}$
5. See student's work.
   Long's Hardware Store $75.10
   Seventy-five and $\frac{10}{100}$
   Answers for name and date will vary.

Page 87

1.

Balance	
891	12
876	55
1352	35

2.

Deposit/Credit		Balance	
		922	07
		887	28
650	25	650	25
		1537	53

Page 88

1. 13 check numbers
2. 6 orders

Page 89

3. 5 checks
4. 12 check numbers
5. 3 total costs
6. 12 orders
7. 3 amounts
8. 7 amounts

Page 90

	Time in Years	$p \times r \times t$ = Interest
1.	$\frac{1}{2}$ or 0.5	$500 × 0.07 × 0.5 = $17.50
2.	1	$1000 × 0.05 × 1 = $50.00
3.	$\frac{1}{3}$ or 0.33	$900 × 0.02 × $\frac{1}{3}$ = $6.00
4.	$\frac{3}{4}$ or 0.75	$600 × 0.035 × 0.75 = $15.75
5.	3	$1600 × 0.04 × 3 = $192.00
6.	5	$6000 × 0.0675 × 5 = $2025.00

7. Change 9 months to $\frac{3}{4}$ or 0.75 year.

Principal		Rate		Time		Interest
p	×	r	×	t		I
$400	×	0.03	×	0.75	=	$9.00

Willis will earn $9.00.

8. $10.00; $810.00
9. $1.50   10. $4.17; 2 years

Page 91

2. Interest for first year:

$p \times r \times t = I$

$\$250 \times 0.05 \times 1 = \$12.50$

Account balance at end of first year:

Prior balance + Interest = New balance

$250 + $12.50 = $262.50

Interest for second year:

$\$262.50 \times 0.05 \times 1 = \$13.13$

George will get $13.13 interest for the second year.

2. $337.08 3. $605.00 4. $191.01
5. $10.82 6. $525.31

Page 92

	Effective Annual Rate	Annual Interest
1.	5.13% or 0.0513	$1000 × 0.0513 = $51.30
2.	2.02% or 0.0202	$5000 × 0.0202 = $101.00
3.	7.25% or 0.0725	$850 × 0.0725 = $61.63
4.	3.05% or 0.0305	$1130 × 0.0305=$34.47
5.	8.32% or 0.0832	$280 × 0.0832 = $23.30
6.	4.08% or 0.0408	$585 × 0.0408 = $23.87

7. $50.00 8. $1.30

Page 93

1. 1.025; $205 2. 1.052; $368.20
3. $630.57 4. $122.95

Page 94

1. $\$200 \times \frac{1}{2} = \$100$

$\$100 \times 4 = \$400$

Alex pays $400 for the bonds.

2. $187.50
3. $62.12 4. $336.00
5. $2258 6. Series EE bonds.

Page 95

	Time in Years	p × r × t = Interest
1.	$\frac{1}{2}$ or 0.5	$5000 × 0.0268 × 0.5 = $67.00
2.	3	$1500 × 0.0328 × 3 = $147.60
3.	$\frac{1}{4}$ or 0.25	$3500 × 0.0289 × 0.25 = $25.29
4.	3	$7200 × 0.0304 × 3 = $656.64
5.	5	$2000 × 0.0523 × 5 = $523.00
6.	1	$2500 × 0.0475 × 1 = $118.75

7. Principal × Rate × Time = Interest

$\$5000 \times 0.031 \times 0.5 = \$77.50$

Jose will receive $77.50 interest.

8. $39.75 9. $78.60; $6078.60 10. $30.40

Page 96

1. $\$5.25 \times 100 = \$525$ in all 2. $450.00
3. $\frac{50}{5000}$ or $\frac{1}{100}$ 4. $254.00
5. $29.00 6. $3.50

Page 97

Time in Years	p × r × t = Interest
1	$500 × .07 × 1 = $35.00
$\frac{1}{2}$ or 0.5	$\$1000 \times 0.07 \times \frac{1}{2} = \$35.00$
$\frac{1}{3}$ or 0.33	$\$800 \times 0.09 \times \frac{1}{3} = \$23.76$
$\frac{1}{4}$ or 0.25	$\$500 \times 0.06 \times \frac{1}{4} = \$7.50$
1	$1200 × 0.055 × 1 = $66.00
$\frac{3}{4}$ or 0.75	$\$900 \times 0.052 \times \frac{3}{4} = \$35.10$

7. Principal Rate (18%) Time ($\frac{6}{12}$) Interest

$\$250 \times 0.18 \times \frac{1}{2} = \$22.50$

Kathy paid $22.50 in interest.

8. $81.25 9. $536.67 10. $6240.00

Page 98

1. Answers will vary.

$p = \$865$, $r = 5.8\%$ or 0.058, $t = \frac{1}{2}$ or 0.5

Round $865 to $900 and 5.8% to 6%.

Principal Rate Time Interest

$\$900 \times 0.06 \times 0.5 = \$27$

Norman will earn about $27.

2. $600 3. $90 4. $800

Page 99

Answers will vary. Accept reasonable estimates.

5. $1400 6. 14 years
7. $100 8. $48
9. $400 10. $600
11. $80 12. $720

Page 100

	Payment			Payment
1.	$17.16		5.	$20.72
2.	$69.44		6.	$56.23
3.	$82.44		7.	$152.66
4.	$85.40		8.	$294.49

9.

Cost		Down payment		Amount of loan
$150	−	$50	=	$100

Payment per $100		Loan in hundreds		Monthly payment
$9.07	×	1	=	$9.07

10. $276.16

Page 101

	Total of Payments	Total Interest
1.	$208.32	$8.32
2.	$533.40	$33.40
3.	$1245.42	$145.42
4.	$2426.04	$126.04

5. $527 6. $3000
7. $162 8. $270; installment loan

Page 102

1.

Old balance		Payment		Balance after payment
$258	−	$50	=	$208

He will be charged a finance charge on $208.

2. $4.16 3. $8.41 4. $274.05
5. $449.45 6. $1744.20

Page 103 Unit 4 Review

1. $3900 2. $110
3. $535.95 4. $31.20
5. $1348.32 6. $270.96
7. $420 8. $715
9. $116.40 10. $9.00

UNIT 5

Page 104

1. Option costs: $599 + $915 = $1514
Sticker price: $8980 + $1514 + 378 = $10,872
The sticker price is $10,872.

2. $11,520 3. $1300 4. $13,686
5. $10,150 6. $14,447

Page 105

1. Dealer's cost:
78% of base price = 0.78 × $9588 = $7478.64
Destination charge = 285.00
$7763.64
To the nearest dollar, the dealer's cost is $7764.

2. $2109 3. $13,972 4. $9941
5. $15,359 6. $11,419

Page 106

1. Accord $2750 Bronco $3750
Camaro $2850 Ciera $2625
Mustang $2650 Shadow $1125

2. Bronco

3. Accord $12,240 Bronco $18,120
Camaro $12,240 Ciera $10,650
Mustang $10,860 Shadow $10,140

4. Accord $11,220 Bronco $16,610
Camaro $11,220 Ciera $9762.50
Mustang $9955 Shadow $9295

Page 107

	Rate	Basic Charge	Chargeable Miles	Additional Charge	Rental Cost
1.	$26.95	$53.90			$53.90
2.	199.00	398.00			398.00
3.	45.95	229.75	274	54.80	284.55
4.	45.95	137.85	314	62.80	200.65
5.	239.00	239.00	450	90.00	329.00

6. 16 days = 2 weeks and 2 days
Basic charge:
2 × \$169.00 + 2 × \$26.95
\$338 + \$53.90 = \$391.90

Basic charge	Sales tax	Rental cost
\$391.90	+ \$391.90 × 0.08	
\$391.90 +	\$31.35	= \$423.25

The total charge is \$423.25.

7. \$161.70; no

Page 108

	Payment
1.	\$117.92
2.	\$68.40
3.	\$160.58
4.	\$61.56

	Payment
5.	\$116.20
6.	\$191.97
7.	\$121.38
8.	\$234.07

9.

Cost	Down Payment	Amount financed
\$11,385 –	\$1385	= \$10,000

The payment is \$2.27 per \$100. So, the monthly payment for a \$10,000 loan is 100 × 2.27 or \$227.00.

10. \$210.40 11. \$332.00 12. \$197.25

Page 109

	Monthly Payment	Total of Payments	Total Interest
1.	\$227.00	\$13,620.00	\$3620.00
2.	\$233.00	\$13,980.00	\$3980.00
3.	\$277.60	\$9,993.60	\$1993.60
4.	\$190.40	\$11,424.00	\$3424.00
5.	\$131.32	\$6,303.36	\$1403.36
6.	\$179.85	\$6,474.60	\$974.60

7.

Total interest Exercise 2	Total interest Exercise 1	
\$3980.00 –	\$3620.00	= \$360

If the interest rate is 13% instead of 14%, \$360 less interest will be paid.

8. \$1430.40 9. \$341.90
10. \$16,411.20; \$17,706.20

Page 110

1. \$121.60 2. \$222.40
3. \$182.60 4. \$140.70
5. \$216.00 6. \$210.00
7. \$227.40 8. \$220.10

Page 111

	Rating Factor	Total Annual Premium
1.	1.2	\$220.68
2.	4.1	\$851.57
3.	2.0	\$304.80
4.	1.7	\$236.39

5.

Bodily injury with limit of 50/100	\$145.20
Property damage with limit of \$50,000	59.30
Collision with \$250 deductible— type B car	111.60
Comprehensive with \$100 deductible— type B car	+ 47.30
	\$363.40

Marilyn's driver-rating factor is 2.0.
2.0 × \$363.40 = \$726.80
The yearly premium for the policy is \$726.80.

6. \$1096.50

Page 112

	Miles Traveled Since Last Purchase	Miles per Gallon	Cost per Mile (in cents)
1.	310	23	6.3¢
2.	407	26	5.8¢
3.	257	23	6.6¢

4. 780 miles ÷ 26 gallons = 30 mpg
5. 23 mpg 6. \$29.52 7. 4.5¢

Page 113

1. 2245 miles 2. 3265 miles
3. 2115 miles 4. 670 miles
5. 1770 miles 6. 1050 miles
7. 21 hours 8. 37.5 gallons
9. \$55.50 10. 1749 miles
11. 58.3 gallons 12. \$89.20
13. 1426 miles

Page 114

1.

Original price	New price	Difference
? –	\$2300 =	\$599

To find the original price, work backwards.
\$599 + \$2300 = \$2899
The original price was \$2899.

2. \$1473 3. \$6816 4. 22¢ or \$0.22

Page 115

5. \$32.95 6. \$4500
7. \$83.40 8. \$216.00
9. 34,572 10. \$1.48

Page 116

1. 815.2
2. 27.84
3. 43.5
4. 76.9
5. $7623
6. $8401
7. 24.15 mpg
8. 25,222

Page 117

	Year 1 30%	Year 2 15%	Year 3 12%	Year 4 8%	Approx. Trade-In Value
1.	$3000	$1500	$1200	$800	$3500
2.	$3729	$1864.50	$1491.60	$994.40	$4350.50
3.	$2946	$1473	$1178.40	$785.60	$3437.00
4.	$4140	$2070	$1656	$1104	$4830
5.	$2022	$1011	$808.80	$539.20	$2359

6. There are 12 months in a year.
   $3000 ÷ 12 = $250
   The average depreciation per month is $250.

7. $130.00

Page 118

1. Totals, 2.25; $30.95
   Labor charge = $101.25
   Total charge = $132.20

2. Labor charge:
   Number of hours × $43.00
   3.25 × $43.00 = $139.75
   The labor charge is $139.75.

3. $13.90
4. $159.50

Page 119

1. $335.81; $396.45; $326.99
2. Shop C has the lowest estimate.
   Shop B has the highest estimate.

Shop B − Shop C
$396.45 − $326.99 = $69.46
Shop C is $69.46 less than Shop B.

3. Shop A; $77.22
4. $5.72
5. Shop C; $218.38
6. $83.66
7. $590

Page 120

1. Single-ride ticket $1.20 × Days per month 24 × Rides per day 2 = Total cost $57.60

   Single-ride tickets $57.60 − Monthly ticket $42.00 = $15.60
   A monthly ticket will cost $15.60 less.

2. less; $3.60
3. $45.00
4. $2.90
5. $12.00
6. $8.50; $3.50

Page 121

1. Flight 172:
   10:16 − 9:10 = 1:06 or 1 hour 6 minutes
   Flight 329:
   9:40 − 7:45 = 1:55 or 1 hour 55 minutes

   55 minutes − 6 minutes = 49 minutes
   Flight 172 takes 49 minutes less time.

2. 2:40 P.M.
3. 4:53 P.M.
4. 4 hours, 30 minutes
5. 2 hours, 16 minutes
6. 4 stops
7. 62
8. 6:11 P.M.
9. 1 hour, 6 minutes
10. 49 minutes

Page 122

1.

Customer	Cost	Miles	Cost per mile
A	$14.40	60	$14.40 ÷ 60 = $0.24
B	$19.20	80	$19.20 ÷ 80 = $0.24

At $0.24 per mile, a customer would pay 70 × $0.24 or $16.80 for 70 miles.

2. $310.80

Page 123

3. Change $179 to $199.
4. $290.70
5. $360
6. Change $29.90 to $20.99.
7. Car E, 20
   Car F, 12.5
8. Change $353.50 to $333.50.

Page 124

	Yearly Expenses	Monthly Expenses
1.	$1425	$118.75
2.	$648	$54.00
3.	$780	$65.00
4.	$360	$30.00
5.	$65	$5.42

	Yearly Expenses	Monthly Expenses
6.	$3278	$273.17
7.	$3600	$300.00
8.	$570	$47.50
9.	$1248	$104.00
10.	$240	$20.00
11.	$75	$6.25
12.	$5733	$477.75

13. 29.8¢
14. 44.1¢

Page 125 Unit 5 Review

1. $9475
2. $495.00
3. $10,908; $1908
4. $731.12
5. 244; 21.4 mpg
6. $34.50
7. $4305
8. $61.45
9. $5.60
10. $3712

UNIT 6

Page 126

	Down Payment	Mortgage Loan
1.	$16,100	$64,400
2.	$30,000	$90,000
3.	$13,500	$76,500
4.	$38,500	$71,500

5. $400 6. $380

Page 127

1.

Amount
$800
$120
$400
$400
$1915

2.

Amount
$1200
$600
$1200
$3485

3.

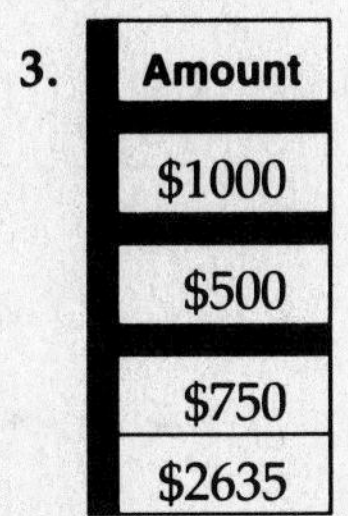

Amount
$1000
$500
$750
$2635

4.

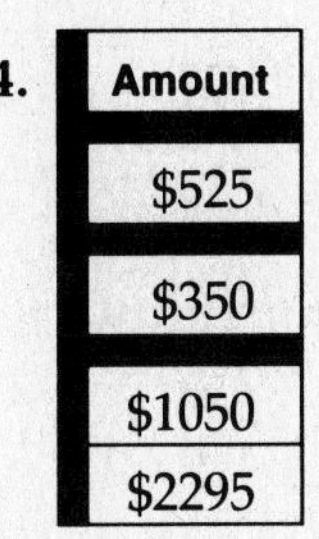

Amount
$525
$350
$1050
$2295

5. No 6. Yes

Page 128

1. The payment is $6.00 per $1000.
   Amount of the loan in thousands:
   $120,000 ÷ $1000 = 120
   120 × $6.00 = $720
   Joyce's monthly payment is $720.

2. $648.90; $489.30
3. $215,460
4. $125,460
5. $145,638; $69,822
6. $15,120

Page 129

	Monthly Payment (Fixed)	Monthly Payment (Adjustable)	Monthly Difference
1.	$332.50	$300.00	$32.50
2.	$587.20	$505.60	$81.60
3.	$805.00	$699.00	$106.00
4.	$922.80	$798.00	$124.80
5.	$419.40	$360.00	$59.40

6. Amount of the loan in thousands:
   $80,000 ÷ 1000 = 80
   Adjustable mortgage: 80 × $6.00 = $480.00
   Fixed-rate mortgage: 80 × $7.34 = $587.20
   $587.20 − $480.00 = $107.20
   Their monthly payment is $107.20 less.

7. $1286.40 8. $532.00 9. $336.00

Page 130

1. 67.5 KWH 2. 7.56 KWH

	Monthly Cost
3.	$0.16
4.	$6.64
5.	$32.00
6.	$0.32
7.	$2.96
8.	$3.68
9.	$7.84
10.	$8.72
11.	$0.96
12.	$0.24

Page 131

1. 2636 KWH
2. 6952 KWH
3. 328 CCF
4. 744 CCF
5. 264 KWH; $21.12
6. $83.62

Page 132

	Annual Premium
1.	$13.60
2.	$17.50
3.	$36.00
4.	$30.00

5. Value of violin in hundreds: $400 ÷ $100 = 4
   Cost of insuring the violin: 4 × $2.50 = $10.00
   Total insurance cost: $340 + $10 = 350
   Marge's total premium is $350.

6. $360.00 7. $528.00 8. $8.00

Page 133

1.
First month's rent		Security deposit		Utilities	
$725	÷	$300	+	$90	= $1,115.00

Sharon will spend $1,115 to rent the apartment the first month.

2. $525.00 3. $411.00 4. $355.00
5. with heat 6. $5.00

Page 134

1. (A) PITI = 26% × gross income
= 0.26 × $2200
= $572
Colleen can afford to spend $572 each month.

2. $578

Page 135

3. $814 4. $1066
5. $44,000 6. $104,000
7. $56,000 8. 0.5 year or 6 months
9. $865 10. $937

Page 136

1. $P = 2l + 2w$
$= 2(8) + 2(6)$
$= 16 + 12$
$= 28$ meters
28 meters − 0.5 meter = 27.5 meters
Andrew will need 27.5 meters of fence.

2. 120 feet 3. 2400 square feet
4. 135 square feet

Page 137

5. $350 6. 375 square meters
7. 358 meters 8. 94 feet
9. 252 square feet 10. $2520.00
11. $500.00 12. $12
13. 48 feet 14. $432.00

Page 138

1.
Length		Width		Height		Volume
40	×	30	×	10	=	12,000 cubic feet

12,000 cubic feet of dirt were removed.

2. 1272 square inches
3. 1080 cubic feet 4. 56 loads
5. 10.125 or $10\frac{1}{8}$ cords 6. 10,400 square feet

Page 139

1. 2 2. 3 3. 3 4. 1
5. 1 6. 4 7. 2 8. 1
9. 62 sq m 10. 0.7 sq m
11. 3350 sq ft 12. 1400 sq in.
13. 40 sq cm 14. 6730 sq mi

Page 140

1. Area of ceiling: 40 × 60 = 2400 square feet
Number of packages needed:
2400 ÷ 64 = 37.5, 38 packages
Cost: 38 packages × $55 = $2090
The estimated cost for the insulation is $2090.

2. $120 3. $1950 4. $2400

Page 141

1. Area of floor: 6 × 8 = 48 square feet
Number of cartons: 48 ÷ 40 = 1.2
Tom will need 2 cartons.
2 × 35 = $70
The tiles will cost $70.00.

2. $660.00 3. 320 tiles
4. $61.00 5. $168.70

Page 142

1.
20 ft
16 ft

2. 10 pieces

Page 143

3.
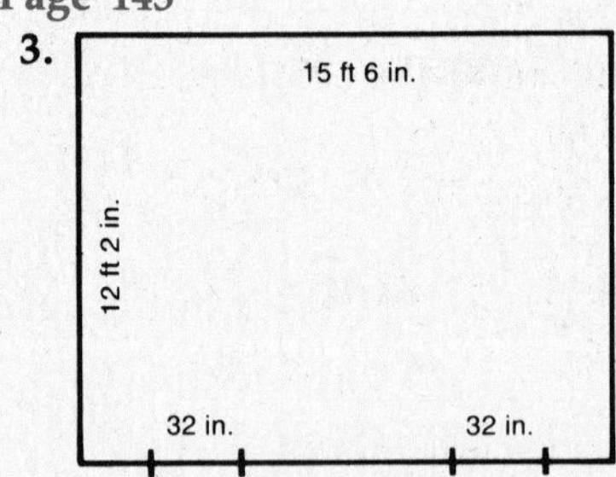

4. 50 feet

5.
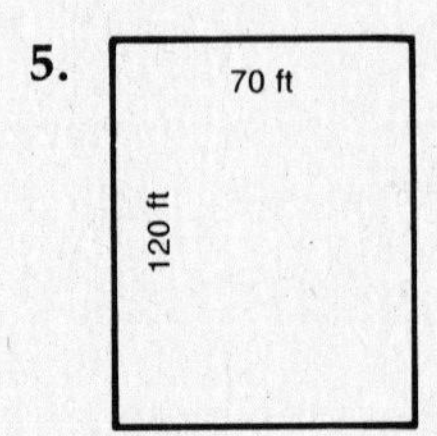

6. 108 feet

7.

70 ft

120 ft

26 ft

30 ft

Street

64 feet

8. 13 feet

9.

A B E D C

10 10 10 10 4

10. 20 square units

11. 100 square units; 120 square ft

Page 144

1.

Dimensions	Area
14 ft × 12 ft	168 sq ft
14 ft × 12 ft	168 sq ft
18 ft × 20 ft	360 sq ft
12 ft × 15 ft	180 sq ft
9 ft × 12 ft	108 sq ft
7 ft × 10 ft	70 sq ft
	1054 sq ft

2. The total area of the house is 1054 square feet.
1054 × $70 = $73,780
The construction cost would be $73,780.

3. $3689 4. $1050 5. $2400
6. $86.40 7. $22.80

Page 145

1. Find 5% of $1350.
0.05 × $1350 = $67.50
Janet saved $67.50.

2. $106.25 3. $89.50 4. $460
5. 11% 6. $440
7. $680.00 8. $23.00

Page 146

1.

	Painting	Vinyl siding
Number of times:	20 ÷ 5 = 4	1
Cost:	4 × $1200 = $4800	$4500

$4800 − $4500 = $300
Bonita will save $300 by buying siding.

2. Brand A; $0.008 per sq ft
3. Brand B 4. Brand B

Page 147 Unit 6 Review

1. $97,500 2. $573.60
3. $59.43 4. $560
5. $1410 6. $500
7. 18,000 cubic feet 8. $2700
9. $96.00 10. 10%

UNIT 7

Page 149

1. $33 2. $61
3. $28 4. $37
5. $22 6. $52
7. $43 8. $20
9. $52 10. $25
11. $34; No
12. $10.77 or $11.00
13. $51
14. $4 more

Page 150

1. 7.65% of $514.56 = 0.0765 × $514.56
= $39.36384
= $39.37

2. $15.52 3. $26.59 4. $61.09
5. $2524.50 6. $6540.75
7. $5713.17 8. $1912.50

Page 151

	Total
1.	$25,290
2.	$12,140
3.	$34,586
4.	$51,172

5. $834 + $490 = $1324
$1324 was withheld for state and local tax.

6. 4%

Page 152

5. owe $327

Page 153

Address information should match W-2 on page 144.

Your social security number: 111 | 00 | 2233

Yes No

	Dollars	Cents
1.	24,893	.00
2.	148	.00
3.	25,041	.00
4.	3,000	.00
5.	22,041	.00
6.	2,000	.00
7.	20,041	.00
8.	3,567	.00
9.	3,287	.00
10.	280	.00
11.		

Page 154

	2. A. Rosner	3. B. Vincent	4. M. Riga	5. C. Sawada
Line 3	4589	22,112	16,489	755
Line 5	1589	19,112	13,489	255
Line 7	1589	17,112	11,489	255
Line 9	238	2567	1723	38
Line 10	$26	$124		
Line 11			$141	$38

6. $13,394
   $2009
   $29
7. $13,811
   $2072
   $221

Page 155

1. 65.8 + 64.4 − 30.38 + 11.98 = 111.8
2. 10,784 + 993 − 6880 − 2399 = 2498
3. 491 − 274 + 1440 + 1950 = 3607
4. 76.92 − 18.6 − 2.26 + 3.68 = 59.74
5. 458.77 + 86.85 + 78.88 + 12.59 = 637.09

Page 156

1. $2816
2. $4886
3. $3716
4. $3784
5. $7353
6. $8860
7. $10,658
   Owe $158
8. $5029
   Refund $196

Page 158

1. Guess 1:
   Account 1 $200
   Account 2 $175
   Difference $25

   Check:
   The total earnings are $375 but the difference is less than $35.

   Guess 2:
   Account 1 $205
   Account 2 $170
   $375

   Check:
   Both the total earnings and the difference are correct.

   Linda received $205 interest from one account and $170 interest from the other.

2. $6100; $9800

Page 159

3. $16,900; $33,800
4. $231; $693
5. $69
6. $65
7. $67
8. A, $17; B, $12; C, $6

Page 161

	Line		Box	Amount
Figure your total income	7	Wages, salaries, tips, etc. This should be shown in box 1 of your W-2 form(s). Attach Form(s) W-2.	7	44718 00
	8a	**Taxable** interest income (see page 25). If over $400, also complete and attach Schedule 1, Part I.	8a	263 00
	b	**Tax-exempt** interest. DO NOT include on line 8a. 8b 130 00		
Attach Copy B of your Forms W-2 and 1099-R here.	9	Dividends. If over $400, also complete and attach Schedule 1, Part II.	9	301 00
	10a	Total IRA distributions. 10a — 10b Taxable amount (see page 26).	10b	
If you didn't get a W-2, see page 24.	11a	Total pensions and annuities. 11a — 11b Taxable amount (see page 26).	11b	
	12	Unemployment compensation (see page 30).	12	
If you are attaching a check or money order, put it on top of any Forms W-2 or 1099-R.	13a	Social security benefits. 13a — 13b Taxable amount (see page 30).	13b	
	14	Add lines 7 through 13b (far right column). This is your **total income.** ▶	14	45282 00
Figure your adjusted gross income	15a	Your IRA deduction (see page 32). 15a 2000 00		
	b	Spouse's IRA deduction (see page 32). 15b 2000 00		
	c	Add lines 15a and 15b. These are your **total adjustments.**	15c	4000 00
	16	Subtract line 15c from line 14. This is your **adjusted gross income.** If less than $23,050 and a child lived with you, see page 63 to find out if you can claim the "Earned income credit" on line 28c. ▶	16	41282 00

Page 163

	Line		Box	Amount
Figure your standard deduction, exemption amount, and taxable income	17	Enter the amount from line 16.	17	41282 00
	18a	Check if: ☐ **You** were 65 or older ☐ Blind; ☐ **Spouse** was 65 or older ☐ Blind. **Enter number of boxes checked** ▶	18a	
	b	If your parent (or someone else) can claim you as a dependent, check here ▶	18b ☐	
	c	If you are married filing separately and your spouse files Form 1040 and itemizes deductions, see page 36 and check here ▶	18c ☐	
	19	Enter the **standard deduction** shown below for your filing status. **But if you checked any box on line 18a or b,** go to page 36 to find your standard deduction. **If you checked box 18c,** enter -0-. • Single—$3,700 • Head of household—$5,450 • Married filing jointly or Qualifying widow(er)—$6,200 • Married filing separately—$3,100	19	6200 00
	20	Subtract line 19 from line 17. If line 19 is more than line 17, enter -0-.	20	35082 00
	21	Multiply $2,350 by the total number of exemptions claimed on line 6e.	21	11750 00
	22	Subtract line 21 from line 20. If line 21 is more than line 20, enter -0-. This is your **taxable income.** ▶	22	23332 00
Figure your tax, credits, and payments	23	Find the tax on the amount on line 22. Check if from: ☐ Tax Table (pages 50–55) or ☐ Form 8615 (see page 38).	23	3499 00
	24a	Credit for child and dependent care expenses. Complete and attach Schedule 2. 24a 960 00		
	b	Credit for the elderly or the disabled. Complete and attach Schedule 3. 24b		
If you want the IRS to figure your tax, see the instructions for line 22 on page 37.	c	Add lines 24a and 24b. These are your **total credits.**	24c	960 00
	25	Subtract line 24c from line 23. If line 24c is more than line 23, enter -0-.	25	2539 00
	26	Advance earned income credit payments from Form W-2.	26	
	27	Add lines 25 and 26. This is your **total tax.** ▶	27	2539 00
	28a	Total Federal income tax withheld. If any tax is from Form(s) 1099, check here. ▶ ☐ 28a 3013 00		
	b	1993 estimated tax payments and amount applied from 1992 return. 28b		
	c	**Earned income credit.** Complete and attach Schedule EIC. 28c		
	d	Add lines 28a, 28b, and 28c. These are your **total payments.** ▶	28d	3013 00
Figure your refund or amount you owe	29	If line 28d is more than line 27, subtract line 27 from line 28d. This is the amount you **overpaid.**	29	474 00
	30	Amount of line 29 you want **refunded to you.**	30	474 00
	31	Amount of line 29 you want **applied to your 1994 estimated tax.** 31		
	32	If line 27 is more than line 28d, subtract line 28d from line 27. This is the **amount you owe.** For details on how to pay, including what to write on your payment, see page 42.	32	
	33	Estimated tax penalty (see page 43). Also, include on line 32. 33		

Page 164

Line	1. Bernard	2. Nguyen	3. Robinson	4. Montoya	5. Kanui
20	33,985	20,780	41,010	52,380	26,310
21	9,400	2,350	7050	4700	2350
22	24,585	18,430	33,960	47,680	23,960
23	3,686	2764	5665	8552	4315
27	2,606	2764	4961	8552	4315
29	132		589		
30		119		1294	735

6. Total income $42,634
   Subtract IRA contributions − 4 000
   Adjusted gross income 38,634
   Subtract standard deduction − 6 200
   Multiply $2350 by 4 and subtract. 32,434
   − 9 400
   Taxable income 23,034
   Their taxable income is $23,034.

7. $2980 8. Refund, $170 9. $29,625

Page 165

	Assessed Value	Assessed Value ÷ 1000	Yearly Real Estate Tax	Monthly Real Estate Tax
1.	$17,400	17.4	$800.40	$66.70
2.	$31,600	31.6	$1738.00	$144.83
3.	$16,800	16.8	$638.40	$53.20
4.	$43,650	43.65	$1178.55	$98.21
5.	$22,750	22.75	$364.00	$30.33
6.	$60,750	60.75	$2430.00	$202.50
7.	$4200	4.2	$214.20	$17.85
8.	$79,600	79.6	$3820.80	$318.40

Page 166

	Total	Should the person itemize?
1.	$4662	No
2.	$7108	Yes
3.	$5508	No
4.	$2367	No
5.	$13,844	Yes
6.	$6957	Yes

7. $879.50

Page 167

1. $8075 + $723 + $12,872 = $21,670
2. $4486
3. $24,840
4. Tax liability $4078
   Refund $1242
5. $32,675
6. Tax liability $4531
   Owe $83

Page 168

1. Guess and Check

   Guess 1:
   Mortgage interest $5000
   Real estate taxes $3400
   Difference $1600

   Check:
   The total is $8400 but the difference is less than $3600.

   Guess 2:
   Mortgage interest $6000
   Real estate taxes $2400
   Difference $3600

   Check:
   Both the total and the difference are correct.

   Kingo paid $6000 in mortgage interest.

2. Work Backwards
   $14\frac{1}{2}$ hours
3. Choose an Operation
   $2216
4. Make a List
   11 numbers
5. Work Backwards
   $5200
6. Make a Drawing
   6 work areas

Page 169

7. Make a table

1040EZ forms	$$	1040A forms	$$	Total $$
1	$8	1	$15	$23
2	$16	1	$15	$31
3	$24	1	$15	$39
4	$32	1	$15	$47
1	$8	2	$30	$38
2	$16	2	$30	$46
3	$24	2	$30	$54

3 1040EZ forms were completed.

8. Make a List
   24 orders
9. Use Estimation
   $1500
10. Guess and Check
    $1600
11. Choose an Operation
    $1933
12. Use Estimation
    $890
13. Find a Pattern
    Change 5853 to 5863
14. Find a Pattern
    Change 1112 to 1121 or 1122

Page 170

	Sales Tax	Total Cost
1.	$12.00	$212.00
2.	$1.04	$14.02
3.	$6.19	$94.64
4.	$0.88	$10.61
5.	$2.46	$40.28
6.	$22.75	$307.10
7.	$813.75	$11,663.75
8.	$0.07	$0.93

9. Total cost: $5.75 + $15.34 + $42.99 = $64.08
Sales tax: 6% of $64.08 = 0.06 × $64.08
= $3.8448 or $3.84
The sales tax is $3.84

10. $92.55

Page 171

1. Total cost: 18 × $1.32 = $23.76
Excise tax: 18 × $0.27 = $4.86

2. 13.4% **3.** $3.06 **4.** 29%
5. $2.79 **6.** 93.2¢

Page 172 Unit 7 Review

1. $164.25 **2.** 2%
3. $17,785; $2667.75 **4.** $29,708
5. Owe $247 **6.** $1123.20
7. $6863; itemize **8.** $5518
9. $7.84; $128.32 **10.** 21.9%

FINAL REVIEW

Page 173

	a	b	c	d	e
1.	4673	$69.60	2.99	15,777	36.39
2.	40,857	$17.57	11.217	19,001	6869
3.	504	9.84	$112.35	7.92	$459.77
4.	70	$7.04	0.8025	55.4	2.6

	a	b	c
5.	4500	651,000	3800
6.	96	4 ft 4 in.	30

	a	b	c	d
7.	$1\frac{2}{3}$	$\frac{1}{6}$	1	$\frac{1}{16}$
8.	$\frac{1}{45}$	$1\frac{1}{6}$	4	$4\frac{4}{9}$
9.	$\frac{29}{30}$	$6\frac{3}{20}$	$7\frac{3}{10}$	$3\frac{13}{16}$
10.	$\frac{7}{24}$	$5\frac{4}{9}$	$3\frac{3}{4}$	$3\frac{71}{72}$

Page 174

	a	b
11.	482	44%
12.	900	99

13. $273.52 **14.** $33,000
15. $468.00 **16.** $412.30
17. $20,280 **18.** $1.45
19. $2760 **20.** $909.11
21. $288.00 **22.** $282.00

Page 175

23. $324.88 **24.** $8.10
25. $10,400 **26.** $116
27. $588.48 **28.** $40.60
29. $5995 **30.** $24.00
31. $605.55 **32.** $50.80

Page 176

33. $455.00 **34.** $3003
35. $135 **36.** 6%
37. $149.33 **38.** $15,508; $2326.20
39. Owe $119 **40.** $1305.60
41. $7154 **42.** $3.09; $59.30

MASTERY TEST

Page 177

	a	b	c	d	e
1.	3453	$33.52	8.52	11,077	36.39
2.	18,878	$29.85	22.269	59,025	6869
3.	2214	9.56	$203.19	4.44	$459.77
4.	60	$3.09	0.905	29.1	4.4

	a	b	c
5.	1700	72,000	3900
6.	64	3 ft 5 in.	48

	a	b	c	d
7.	$1\frac{1}{3}$	$\frac{3}{5}$	$\frac{1}{3}$	3
8.	$\frac{1}{50}$	$1\frac{1}{4}$	6	$4\frac{1}{6}$
9.	$\frac{11}{12}$	$7\frac{11}{12}$	$4\frac{7}{8}$	$5\frac{3}{10}$
10.	$\frac{13}{30}$	$4\frac{4}{7}$	$7\frac{1}{3}$	$3\frac{19}{20}$

Page 178

	a	b
11.	252	32%
12.	300	45

13. $304.00 **14.** $28,964
15. $581 **16.** $301.36
17. $6084 **18.** $1.30
19. $2800 **20.** $499.31
21. $617.70 **22.** $180

Page 179

23. $211.92 **24.** $13.30
25. $13,042 **26.** $206.00
27. $836.16 **28.** $54.00
29. $5547 **30.** $12.00
31. $682.62 **32.** $29.96

Page 180

33. $612 **34.** $2976
35. $74 **36.** 5%
37. $58.83 **38.** $13,534; $2030.10
39. Refund; $233 **40.** $1271.90
41. $5447 **42.** $3.60; $58.93

MATH SYMBOLS

Symbol	Meaning	Example
$+$	add (plus)	$4 + 7$ 4 plus 7 more is 11
$-$	subtract (minus, take away)	$11 - 7$ 11 minus 7 is 4
$\times$	multiply (times)	3×7 3 times 7 is 21
$\div$	divide	$21 \div 7$ 21 divided by 7 is 3
$=$	equal to	$5 + 1 = 6$ 5 plus 1 is equal to 6
$\approx$	approximately equal to	$0.66 \approx 2/3$ 0.66 is about two-thirds
$\neq$	does not equal	$5 + 1 \neq 7$ 5 plus 1 does not equal 7
$>$	greater than	$6 > 5$ 6 is greater than 5
$<$	less than	$4 < 5$ 4 is less than 5
$\geq$	greater than or equal to	$n \geq 5$ so, *n* is 5 or more
$\leq$	less than or equal to	$n \leq 5$ so, *n* is 5 or less
$^\circ$	degrees	52°C is 52 degrees Celsius 52°F is 52 degrees Fahrenheit
π	pi	pi ≈ 3.14 pi is approximately 3.14
%	percent	5% of 80 is 4 5 percent of 80 is 4
¢	cent	4 pennies equal 4¢ 4 pennies equal 4 cents
$	dollar	100¢ equal $1.00 100 cents equal one dollar

TABLES OF MEASUREMENTS

Time	
60 seconds = 1 minute	52 weeks = 1 year
60 minutes = 1 hour	12 months = 1 year
24 hours = 1 day	365 days = 1 year
7 days = 1 week	
U.S. Customary Length	**Metric Length**
12 inches (in.) = 1 foot (ft)	10 millimeters (mm) = 1 centimeter (cm)
36 inches = 1 yard (yd)	100 centimeters = 1 meter (m)
3 feet = 1 yard	1,000 meters = 1 kilometer (km)
5,280 feet = 1 mile (mi)	
1,760 yards = 1 mile	
U.S. Customary Weight	**Metric Weight (Mass)**
16 ounces (oz) = 1 pound (lb)	1,000 milligrams (mg) = 1 gram (g)
2,000 pounds = 1 ton (t)	1,000 grams = 1 kilogram (kg)
U.S. Customary Capacity	**Metric Capacity**
8 ounces (oz) = 1 cup (c)	1,000 milliliters (mL) = 1 liter (L)
2 cups = 1 pint (pt)	
4 cups = 1 quart (qt)	
2 pints = 1 quart	
4 quarts = 1 gallon (gal)	
8 pints = 1 gallon	
16 cups = 1 gallon	

FORMULAS

Shape	Description	Formula
Perimeter		
	square	$P = 4s$
	rectangle	$P = 2l + 2w$
	circumference of a circle	$C = \pi d$ $C = 2\pi r$
Area		
	square	A = s2
	rectangle	$A = lw$
	parallelogram	$A = bh$
	triangle	$A = \frac{1}{2} bh$
	circle	$A = \pi r^2$
Volume		
	cube	$V = s^3$
	rectangular prism	$V = lwh$
	cylinder	$V = \pi r^2 h$
Temperature		
°C	centigrade	°C = .56(°F − 32) or 5/9(°F − 32)
°F	Fahrenheit	°F = 1.8(°C) + 32 or (9/5 × °C) + 32

OTHER FORMULAS

Formula	Description
Simple Interest	
$I = prt$	Interest is equal to principal × rate × time
Distance	
$d = rt$	Distance is equal to rate × time
Total Cost	
$c = nr$	Total cost is equal to number of units × cost per unit
Electricity	
	1 kilowatt-hour = 1,000 watt-hours amps = watts ÷ volts

INDEX